SAGE

ICRP Publication 112

Annals of the ICRP

Preventing Accidental Exposures from New External Beam Radiation Therapy Technologies

ICRP PUBLICATION 112

Approved by the Commission in September 2009

Abstract–Disseminating the knowledge and lessons learned from accidental exposures is crucial in preventing re-occurrence. This is particularly important in radiation therapy; the only application of radiation in which very high radiation doses are deliberately given to patients to achieve cure or palliation of disease.

Lessons from accidental exposures are, therefore, an invaluable resource for revealing vulnerable aspects of the practice of radiotherapy, and for providing guidance for the prevention of future occurrences. These lessons have successfully been applied to avoid catastrophic events with conventional technologies and techniques. Recommendations, for example, include the independent verification of beam calibration and independent calculation of the treatment times and monitor units for external beam radiotherapy, and the monitoring of patients and their clothes immediately after brachytherapy.

New technologies are meant to bring substantial improvement to radiation therapy. However, this is often achieved with a considerable increase in complexity, which in turn brings opportunities for new types of human error and problems with equipment. Dissemination of information on these errors or mistakes as soon as it becomes available is crucial in radiation therapy with new technologies. In addition, information on circumstances that almost resulted in serious consequences (near-misses) is also important, as the same type of events may occur elsewhere. Sharing information about near-misses is thus a complementary important aspect of prevention. Lessons from retrospective information are provided in Sections 2 and 4 of this report.

Disseminating lessons learned for serious incidents is necessary but not sufficient when dealing with new technologies. It is of utmost importance to be proactive and continually strive to answer questions such as 'What else can go wrong', 'How likely is it?' and 'What kind of cost-effective choices do I have for prevention?'. These questions are addressed in Sections 3 and 5 of this report.

1

Section 6 contains the conclusions and recommendations. This report is expected to be a valuable resource for radiation oncologists, hospital administrators, medical physicists, technologists, dosimetrists, maintenance engineers, radiation safety specialists, and regulators. While the report applies specifically to new external beam therapies, the general principles for prevention are applicable to the broad range of radiotherapy practices where mistakes could result in serious consequences for the patient and practitioner.

Keywords: Accidental exposure; Radiation therapy; New technologies; Retrospective methods; Prospective methods

SAGE

ICRP Publication 112

Annals of the ICRP

Guest Editorial

NEW TECHNOLOGIES, NEW RISKS

ICRP *Publication 86*, 'Prevention of accidental exposures to patients undergoing radiation therapy', was published in 2000 (ICRP, 2000). The usual life span of the ICRP recommendations exceeds, sometimes by far, a full decade. Consequently, it may appear somewhat surprising to see ICRP publishing a new document focusing on the risks of accidents in radiotherapy less than 10 years after ICRP *Publication 86*.

In fact, the authors of ICRP *Publication 86* had somehow anticipated such a need: a few sentences found in the text appear to foreshadow this publication. A full section (5.9) was devoted to 'The potential for accidental exposures in the future'. What was written in this section deserves to be repeated here:

'The recommendations ... [in this publication] are based on a retrospective analysis of accidental exposures in radiation therapy with past and current types of equipment. There are, however, a number of factors that may cause a change in this picture in the future:

- With the worldwide expansion of radiotherapy there may be more accidents related to inadequate staff training ....
- There is a common misperception that modern equipment is safer and will require less quality assurance.
- ...Accidents may occur due to inadequate accelerator maintenance.... The increased number of computer-controlled systems may also lead to more computer related accidents, compared to mechanical failures.
- The new technologies of high dose rate (HDR) brachytherapy, "gamma knife" therapy units, multi-leaf collimators, and intensity modulated radiotherapy (IMRT) may produce new types of accidental exposures.'

Moreover, the Summary of ICRP *Publication 86* notes that 'Major accidental exposures are rare, but it is likely that they will continue to happen unless awareness is increased. Accidents will usually occur as the result of inadequate education and training, lack of quality assurance, poor infrastructure, equipment failure, and improper decommissioning. Unless these issues are properly addressed and dealt with, more accidental exposures are likely to occur, as current and new technologies developments are disseminated.'

Actually, the authors of ICRP *Publication 86* would clearly have preferred to be wrong! Unfortunately, they were not and it has recently become apparent that their pessimistic predictions were partly right.

Obviously, ICRP acknowledges the impressive recent technological developments of modern radiotherapy, primarily 'conformal' radiotherapy and IMRT. It must be recognised that these developments brought new and unmatched precision to radiation oncologists. This precision allows both a reduction in the volume of healthy tissue that has to be irradiated (with a consequential reduction in side effects and complications), and a dose escalation to the tumour, which has already proven in some instances (e.g. prostate cancer) to increase the cure rate significantly.

Unfortunately, the dark side of these successes is that the new and more sophisticated technologies give rise to new types of accidents. Let us take a single, simple example. With so-called 'conventional' or 'classical' radiotherapy (on which ICRP *Publication 86* was primarily based), all data for the treatment of a given patient had to be re-entered for each of the fractions. In this setting, ICRP *Publication 86* noted that, for an entire treatment, approximately 1000 parameters (!) had to be entered manually. This obviously led to a non-negligible risk of human error. However, it has to be emphasised that the errors generally occurred for no more than a few fractions, and usually did not lead to significant 'over' or 'under' doses. In other words, the conventional procedures most probably led to a significant number of errors, but usually with limited deviations from the prescribed total dose, and therefore without serious consequences for the patients.

With the general introduction of computers in radiotherapy, things have changed. Today, all data are registered from the beginning and 'recalled' automatically for each fraction. On one hand, this avoids the risk of deviations due to multiple data re-entries. On the other hand if, unfortunately, some error arises during the initial data entry, it may impact on the entire treatment, with dose deviations which are generally much larger than before. In summary, the new techniques result in far fewer errors, but such errors, when they do occur, can be much more severe.

Moreover, new types of accidents have been encountered due to: the complexity of the present treatment preparations; the increased sophistication of the whole treatment process (with an increasing number of steps and more people involved); the omnipresence of computers with frequent and regular upgrading of more and more complicated software; and the difficulty of regularly and correctly training all the physicians, physicists, dosimetrists, engineers, etc. involved in a busy radiotherapy unit.

Such accidents have been honestly recorded and analysed in various countries. They have been responsible for complications of varying severity, including the deaths of patients in some instances.

ICRP could not ignore or dismiss this reality, so the decision was made to develop a new document dealing with these new risks in modern radiotherapy. Pedro Ortiz López, who had been the Chair of the ICRP Task Group that wrote ICRP *Publication 86*, agreed to form a new Task Group. The role of this new Task Group was to propose new recommendations for the prevention of accidents in radiotherapy, taking into account the changes and the more recent developments of the last decade.

These new recommendations are based on: lessons learned through experience; the continuous control of computer software; regular control of the ballistics; precise checking of the delivered dose (in particular, by in-vivo dosimetry); and more. As such, it is the hope of this Task Group, Committee 3, and ICRP as a whole that these new recommendations will help to reduce the risk of accidents in modern radiotherapy significantly, both in terms of frequency and severity.

The ultimate goal is to be able to offer cancer patients the more recent and sophisticated modern radiotherapy techniques, which are expected to be more efficient and with less chance of side effects, in safer conditions.

JEAN-MARC COSSET

Reference

ICRP, 2000. Prevention of accidental exposures to patients undergoing radiation therapy. ICRP Publication 86. Ann. ICRP 30 (3).

CONTENTS

PREFACE

Over the years, the International Commission on Radiological Protection (ICRP), referred to below as 'the Commission', has issued many reports providing advice on radiological protection and safety in medicine (ICRP *Publication 105* is a general overview of this area). These reports summarise the general principles of radiation protection, and provide advice on the application of these principles to the various uses of ionising radiation in medicine and biomedical research.

Most of these reports are of a general nature, and the Commission wishes to address some specific situations where difficulties have been observed. It is desirable that reports on such problem areas be written in a style which is accessible to those who may be directly concerned in their daily work, and that every effort is taken to ensure wide circulation of such reports.

A first step in this direction was taken at the Commission's meeting in Oxford, UK in September 1997. At that time, on the recommendation of ICRP Committee 3, the Commission established several Task Groups to produce reports on topical issues in medical radiation protection.

Several such reports have already appeared in print (ICRP *Publications 84, 85, 86, 87, 93, 94, 97, 98,* and *102,* and ICRP *Supporting Guidance 2*). The present report continues this series of concise and focused documents, and several more advisory reports are being prepared. ICRP *Publication 86*, published in 2000, dealt with the prevention of accidental exposure of radiation therapy patients. That report contained the lessons learned from real case histories of major accidental exposures, and provided recommendations to prevent re-occurrence. Most of these lessons stemmed from conventional radiation therapy, and little mention was made of newer technologies.

From the time of ICRP *Publication 86* to date, a number of reports on events that have occurred with new technologies have become available. In addition, considerable progress has been made in prospective methods of anticipating what else can happen. At their meeting in Berlin, Germany in October 2007, the Commission launched a Task Group on Preventing Accidental Exposure with New Technologies in Radiation Therapy. Its terms of reference were to re-evaluate which lessons from conventional techniques are still applicable, and to review available reports on case histories with new technologies, as well as prospective methods of prevention.

The membership of the Task Group was as follows:

P. Ortiz López (Chairman)   O. Holmberg
J.M. Cosset   J.C. Rosenwald
P. Dunscombe

The corresponding members were:

L. Pinillos Ashton   J.J. Vilaragut Llanes   S. Vatnitsky

The Task Group met in Paris in June 2008 and prepared an advanced draft, which was approved by Committee 3 in September 2008 and by the Main Commission in October 2008 for its customary public consultation via the Internet. Numerous helpful comments from this consultation have greatly contributed to the final version of the report.

The membership of Committee 3 during the period of preparation of this report was:

C. Cousins, Chair	E. Vañó, Chair	J.M. Cosset
(to October 2008)	(from October 2008)	(Vice Chair)
I. Gusev	J.W. Hopewell	Y. Li
P-L. Khong	J. Liniecki	S. Mattsson
P. Ortiz López	L. Pinillos Ashton	M.M. Rehani (Secretary)
H. Ringertz	M. Rosenstein	Y. Yonekura
B. Yue		

This report aims to serve the purposes described above. In order to be as useful as possible for those purposes, its style differs in a few respects from the usual style of the Commission's publications in the Annals of the ICRP.

The report was approved for publication by the Commission in September 2009.

References

ICRP, 2000a. Pregnancy and medical radiation. ICRP Publication 84. Ann. ICRP 30 (1).

ICRP, 2000b. Avoidance of radiation injuries from medical interventional procedures. ICRP Publication 85. Ann. ICRP 30 (2).

ICRP, 2000c. Prevention of accidental exposures to patients undergoing radiation therapy. ICRP Publication 86. Ann. ICRP 30 (3).

ICRP, 2000d. Managing patient dose in computed tomography. ICRP Publication 87. Ann. ICRP 30 (4).

ICRP, 2001. Radiation and your patient: a guide for medical practitioners. ICRP Supporting Guidance 2. Ann. ICRP 31 (4).

ICRP, 2004a. Managing patient dose in digital radiology. ICRP Publication 93. Ann. ICRP 34 (1).

ICRP, 2004b. Release of patients after therapy with unsealed radionuclides. ICRP Publication 94. Ann. ICRP 34 (2).

ICRP, 2005a. Prevention of high-dose-rate brachytherapy accidents. ICRP Publication 97. Ann. ICRP 35 (2).

ICRP, 2005b. Radiation safety aspects of brachytherapy for prostate cancer using permanently implanted sources. ICRP Publication 98. Ann. ICRP 35 (3).

ICRP, 2007a. Managing patient dose in multi-detector computed tomography (MDCT). ICRP Publication 102. Ann. ICRP 37 (1).

ICRP, 2007b. Radiological Protection in Medicine. ICRP Publication 105. Ann. ICRP 37 (6).

EXECUTIVE SUMMARY

(a) The decision to implement a new technology for radiation therapy should be based on a thorough evaluation of the expected benefits, rather than being driven by the technology itself. To ensure safe implementation, a step-by-step approach should be followed.

(b) ICRP *Publication 86* (2000) concluded that 'purchasing new equipment without a concomitant effort on education and training and on a programme of quality assurance is dangerous'. Although originally referring to conventional radiation therapy, this conclusion is even more critical for new technologies.

(c) Major safety issues can arise from underestimating the staff resources required to implement and operate a new technology. Resources should be allocated in order to avoid the substitution of proper training with a short briefing or demonstration, from which important safety implications of new techniques cannot be fully appreciated.

(d) Certain tasks, such as calibration, beam characterisation, complex treatment planning, and pretreatment verification for intensity modulated radiotherapy (IMRT), require a substantial increase in staff allocation. The re-assessment of staff requirements, in terms of training and the number of professionals, is essential when moving to new technologies.

(e) Radiation therapy staff and hospital administrators should remain cognisant of the fact that the primary responsibility for the safe delivery of treatment lies with them. This responsibility includes investigating discrepancies in dose measurements before using the beam for patient treatments. Independent verification of beam calibration remains essential.

(f) Hospital administrators of radiation therapy departments should provide a work environment that facilitates concentration and avoids distraction.

(g) Manufacturers should be aware of their responsibility for delivering the correct equipment with the correct calibration files and accompanying documents. They also have a responsibility to provide correct information and advice, upon request, to users. Procedures to meet these responsibilities should be developed and maintained in a quality control environment.

(h) Programmes for purchasing, acceptance testing, and commissioning should not only address treatment machines but also treatment planning systems (TPSs), radiation therapy information systems (RTISs), imaging equipment used for radiation therapy, software, procedures, and entire clinical processes. Devices and processes should be recommissioned after equipment modifications, including software upgrades and updates.

(i) Procedures should be in place to deal with situations created by computer 'crashes', which may cause loss of data integrity and lead to severe accidental exposures.

(j) Protocols for treatment prescription, reporting, and recording, such as found in reports of the International Commission on Radiation Units and Measurements (ICRU), should be revised to accommodate new technologies. They should be adopted at a national level with the support of professional bodies. Similarly, dosimetry protocols should be developed for small and non-standard radiation fields.

(k) Target dose escalation without a concomitant increase in the probability of normal tissue complications generally implies a reduction of geometrical margins. Such a reduction is only possible with conformal therapy accompanied by precise, image-guided patient positioning and effective immobilisation, together with a clear understanding of the accuracy achieved in clinical practice. Without these features, target dose escalation could lead to severe patient complications.

(l) Unambiguous, well-structured communication is essential, considering the complexity of radiation therapy and the multidisciplinary nature of the healthcare environment. In particular, procedures to notify physicists of maintenance and repair activities, identified as crucial in conventional technology, are even more important with new technologies.

(m) When conventional tests and checks are not applicable or not effective for new technologies, the safety philosophy should aim to identify measures to maintain the required level of safety. This may require the design of new tests, or the modification and validation of existing tests.

(n) Lessons learned from past accidental exposures should be incorporated into training. Radiation therapy facilities are encouraged to share their experiences of actual and potential safety incidents through participation in databases such as the Radiation Oncology Safety Information System (ROSIS). Report formats and analytical tools should be further developed to maximise and facilitate the learning components of such databases.

(o) Prior to the introduction of new techniques and technologies, there is little or no operational experience to share. To maintain safety in this situation, two complementary measures are recommended:

- Prospective safety assessments should be undertaken in order to develop risk-informed and cost-effective quality assurance programmes. Examples include failure modes and effects analysis, probabilistic safety assessment, and risk matrix.
- Moderated electronic networks and panels of experts supported by professional bodies should be established in order to expedite the sharing of knowledge in the early phase of introducing a new technology.

References

ICRP, 2000. Prevention of accidental exposures to patients undergoing radiation therapy. ICRP Publication 86. Ann. ICRP 30 (3).

1. INTRODUCTION

1.1. Background

(1) New technologies have been introduced into radiation therapy with the principal aim of improving treatment outcome by means of dose distributions which conform more strictly to tumour (clinical target) volumes. A highly conformal dose distribution allows for dose escalation in the target volume without increasing the radiation dose to neighbouring normal tissues, or for a reduction in radiation dose to normal tissues without decreasing the dose to the target. These new technologies encompass the increased use of multileaf collimators (MLCs), IMRT, volumetric arc therapy (VMAT), tomotherapy, image-guided radiation therapy (IGRT), respiratory gating, robotic systems, radiosurgery, newer and more complex treatment planning systems (TPSs), virtual simulation, and 'all-inclusive' electronic patient data management systems.

1.2. Trends in radiation therapy

(2) The development of these new technologies has been impressively fast over the last few years. As an example, 10 years after a new concept for the delivery of dynamic conformal radiotherapy, tomotherapy (Mackie et al., 1993), was described, a machine was put into clinical operation. Five years after the first clinical prototype, more than 200 tomotherapy machines are operating throughout the world. Although most machines are located in North America and Western Europe, a significant number are installed in Asia, and the technology is also reaching the Middle East.

(3) Similar trends are observed for other rotational approaches using computer-controlled intensity modulation. These are being implemented on more 'traditional' accelerators and are classified as VMAT techniques, with the trade name depending on the manufacturer. Other recent developments include online volumetric imaging [cone beam computed tomography (CT) with kV or MV beams], robotic solutions, particle therapy with proton or ion beams, and patient data management systems with 'record and verify' (R&V) capabilities.

(4) There is, therefore, a continuous evolution of what is considered a 'standard' machine for radiotherapy towards more sophisticated equipment which, in turn, requires computer control for its operation. This standard varies widely throughout different parts of the world. For the high-income countries (North America and Western Europe), the standard has become, in the last 5 years, an accelerator equipped with an MLC and a flat panel portal imaging device, integrated with a patient data management system. More recently, an accelerator capable of delivering IMRT has become the standard for many countries.

(5) While a technologically simple cobalt machine, combined with a two-dimensional TPS, was the standard in the low-income countries, there is now a trend in such countries towards more sophisticated solutions, enabled by the acquisition of the necessary equipment. However, there is a danger of being driven by technology

itself, rather than by a sound evaluation of the documented evidence of the expected benefits for patients. There is a further danger of proceeding to implement new technology and techniques without ensuring the availability of sufficient adequately trained technical, scientific, and medical staff, and insufficient human resources to cope with labour-intensive tasks required by new technologies.

1.3. Trends in risk assessment

(6) Minimising the risk of accidental exposures of radiation therapy patients has been based largely on compliance with regulatory requirements, codes of practice, and international standards. These can be considered 'prescriptive approaches'. Retrospective compilations of lessons learned from the review and analysis of accidental exposures in radiation therapy have been published (IAEA, 2000; ICRP, 2000). These can be used to check whether a given radiation therapy department has sufficient provisions in place to avoid accidental exposures similar to those reported. This is a 'retrospective' approach to prevention. As an example, major accidental exposures caused by errors in the calibration and commissioning of radiation therapy equipment have led to putting preventive measures in place, such as an independent redundant determination of the absorbed dose to detect possible beam calibration errors[1].

(7) Focusing on major events with catastrophic consequences and very low probability of occurrence may result in overlooking other types of error that can occur with a higher probability and have lower, but still significant, consequences. Some approaches have been put in place to share 'near-misses' and mistakes with low consequences, but which may have had severe consequences under different circumstances. A good example is the Radiation Oncology Safety Information System (ROSIS) (http://www.rosis.info).

(8) Retrospective approaches are limited to reported experiences, i.e. they do not address what else could go wrong, or identify other potential hazards. Thus, latent risks from other possible failure modes which have not yet occurred or have not been shared or published will remain unaddressed unless more 'prospective approaches' are applied.

(9) Prospective approaches to the identification and analysis of failure modes, assessment of their frequency and consequences, and their evaluation in terms of risk are available and are being used by some healthcare institutions to provide

---

[1] Thermoluminiscence dosimetry (TLD) postal services to audit calibration of radiation therapy units are available from various services, such as the International Atomic Energy Agency's TLD postal audit and the Radiological Physics Center in Houston, TX, USA. In all instances, the services provide an independent and impartial quality audit of the dosimetry at the radiotherapy departments. A postal TLD audit is an auditing method for checking the hospital dosimetry, by which an external laboratory mails a set of thermoluminiscence dosimeters (TLDs) to a radiotherapy department to be audited. The TLDs are irradiated in the radiotherapy department and mailed back to the laboratory, together with a statement on the radiation dose delivered (according to the local dosimetry at the radiotherapy department). The TLDs are then read at the laboratory and the reading values compared with the dose values stated by the radiotherapy department.

risk-informed strategies. Such approaches have started to be adopted by the radiation therapy community (Huq et al., 2007, 2008; Duménigo et al., 2008; Ortiz López et al., 2008; Vilaragut Llanes et al., 2008).

(10) In summary, both retrospective and prospective approaches are needed if the introduction of new technologies and techniques is to enhance the quality of patient treatment without compromising safety.

1.4. Objectives of this report

(11) The objectives of this report are to summarise lessons from experience to date, and also to provide guidance on prospective approaches to the reduction of risk of accidental exposures in radiation therapy, with emphasis on advanced and complex technologies and techniques.

1.5. Scope

(12) This report focuses on approaches to preventing accidental exposures, or mitigating their consequences, with respect to new technologies for external beam radiation therapy. The scope is confined to radiation safety issues, i.e. to identify preventive measures to avoid accidental exposure by retrospective and prospective methods. These measures may affect different aspects of the radiotherapy programme, including quality control, but these aspects are outside the scope of this report. The report uses conclusions and recommendations from ICRP *Publication 86* (ICRP, 2000) devoted to radiation therapy with conventional technologies, but full detailed conclusions and recommendations from that publication are outside the scope of this report. In addition, other ICRP publications are devoted to high-dose-rate brachytherapy and permanent implants, and therefore this report does not deal with these techniques specifically.

1.6. Structure

(13) In Section 2, lessons learnt from conventional radiation therapy are summarised and discussed in the context of their applicability to new technologies being introduced into the clinic. In Section 3, a review of new technologies and their implications for safety are presented. Reported case histories of accidental exposures and near-misses when using new technologies, and the lessons learnt from them, are provided in Section 4. Three prospective approaches to enhancing safety are described in Section 5. These approaches enable the prioritisation of activities aimed at reducing the frequency of occurrence of errors and their severity, and optimising the quality management system so that errors may be detected before they impact on clinical treatment. A recapitulation of lessons learnt and recommendations are given in Section 6. A larger sample of case histories is provided in Annex A.

1.7. References

Duménigo, C., Ramírez, M.L., Ortiz López, P., et al., 2008. Risk analysis methods: their importance for safety assessment of practices using radiation. XII Congress of the International Association of Radiation Protection, IRPA 12, 19–24 October 2008, Buenos Aires, Argentina. Book of Abstracts. Full paper available at: http://www.irpa12.org.ar/fullpaper_list.php.

Huq, S., ASTRO, AAPM, NCI, 2007. A method for evaluating QA needs in radiation therapy. Symposium on Quality Assurance of Radiation Therapy: Challenges of Advanced Technology, 20–22 February 2007, Dallas, TX.

Huq, M.S., Fraass, B.A., Dunscombe, P.B., et al., 2008. A method for evaluating quality assurance needs in radiation therapy. Int. J. Radiat. Oncol. Biol. Phys. 70, S170–S173.

IAEA, 2000. Lessons Learned from Accidental Exposure in Radiotherapy. Safety Report Series No. 17. International Atomic Energy Agency, Vienna.

ICRP, 2000. Prevention of accidental exposures to patients undergoing radiation therapy. ICRP Publication 86. Ann. ICRP 30 (3).

Mackie, T.R., Holmes, T., Swerdloff, S., et al., 1993. Tomotherapy: a new concept for the delivery of dynamic conformal radiotherapy. Med. Phys. 20, 1709–1719.

Ortiz López, P., Duménigo, C., Ramírez, M.L., et al., 2008. Risk analysis methods: their importance for the safety assessment of radiotherapy. Annual Congress of the European Society of Therapeutic Radiology and Oncology (ESTRO 27), 14–17 September 2008, Goteborg. Book of Abstracts.

Vilaragut Llanes, J.J., Ferro Fernández, R., Rodriguez Martı, M., et al., 2008. Probabilistic safety assessment (PSA) of the radiation therapy treatment process with an electron linear accelerator (LINAC) for medical uses. XII Congress of the International Association of Radiation Protection, IRPA 12, 19–24 October, 2008, Buenos Aires. Book of Abstracts. Full paper available at: http://www.irpa12.org.ar/fullpaper_list.php.

2. SUMMARY OF LESSONS FROM ACCIDENTAL EXPOSURES WITH CONVENTIONAL TECHNOLOGY

2.1. Organisation and quality management system

(14) ICRP *Publication 86* (ICRP, 2000) points out that most severe accidental exposures have occurred in radiation therapy departments where a quality assurance[1] programme was either not in place or, if it was, was not fully implemented and/or monitored. Weaknesses identified from accidental exposures with conventional technology were:

- insufficient education and training, including poor understanding of the physics of the treatment equipment and TPSs;
- absence of appropriate acceptance and commissioning procedures;
- misunderstanding of instructions for users;
- reliance on verbally communicated instructions;
- omission of some of the quality control checks;
- changing a procedure without validation;
- resuming treatments after a major repair without notifying the responsible person for dosimetry verification;
- poor notification of unusual tissue reactions; and
- poor patient follow-up.

(15) ICRP *Publication 86* (ICRP, 2000) states that a comprehensive quality assurance programme can lead to the detection of systematic errors and decrease the frequency and severity of random errors. Minimising the probability of occurrence and severity of accidental exposures can be achieved with reasonable effort and expense in a radiation therapy department when 'two conditions are fulfilled: i) a comprehensive and coherent quality assurance programme is in place and ii) some in-vivo dose measurements are performed'.

2.1.1. Recommendation

(16) Hospital managers need to put in place a quality management system that addresses education, training, continuous professional development, assessment of the required number and qualification of staff, appropriate assignment of duties and responsibilities of qualified staff, a clear organisational structure, written procedures, and supervision of compliance. Procedures should encompass equipment purchasing, acceptance testing and commissioning, regular quality control checks, equipment use and maintenance, effective communication throughout the treatment process, patient observation, and follow-up of abnormal tissue reactions. Regular

---

[1] The term 'quality assurance' tends to be replaced by the more comprehensive 'quality management', which includes not only technical issues of radiotherapy but also organisational issues. However, when quoting text from ICRP *Publication 86*, the original terminology is kept for easier reference.

17

re-assessment of the number of staff, and relevant training and competence is of particular importance as the workload increases, new equipment is purchased, and new
techniques are introduced into the radiation therapy programme. The quality management system should include provisions for well-structured quality audits, such as
those described by the International Atomic Energy Agency (IAEA, 2007a,b, 2008a).

2.2. Special problem of the availability of staff with training and competence

(17) In many parts of the world, particularly in low-income countries, the lack of
staff with the training and competence essential for safety remains an unresolved issue. Shortages exist for radiation oncologists, medical physicists, technologists,
dosimetrists, and maintenance engineers. In particular, medical physicists, responsible for safety-critical issues such as calibration of radiation beams and sources, dosimetric treatment planning, and radiation physics aspects of quality control checks,
are unavailable in many countries. The reasons for this shortage are two-fold. First,
there may be no established programme of education and hands-on training for these
professionals. In some countries, it may not be feasible to maintain such an educa-
tional programme at a national level as only a small number of professionals may be
required. Furthermore, in many countries, the profession of medical physics is not
formally recognised; as a consequence, appropriate candidates may not be attracted
to the field. Second, sending professionals abroad for education and hands-on training often results in losing them permanently as they may prefer to stay in the country
where they have been trained. This is particularly true if the profession of medical
physics is not formally recognised in their own countries.

2.2.1. Recommendations

(18) Governments need to be aware of these difficulties when embarking on or
operating a radiation therapy programme. Governments need to make provisions
for a system of education and training (in the country or abroad), have in place a
process of certification for medical physicists, and develop a programme to retain
staff who are essential in safety maintenance.

(19) The general recommendations from conventional radiation therapy summarised above are equally valid for new technologies, although some (e.g. in-vivo dose
measurements) may require specific adaptation for certain procedures (e.g. IMRT).

2.3. Safety culture

(20) A number of reported accidental exposures have been linked to inattention to,
and lack of awareness of, signs which indicated that 'something may be going
wrong', such as conflicting signals, error messages, and equipment malfunctions.
General unawareness of the significance of a situation or fault was a common feature
of many major accidental exposures. As an example of lack of full awareness of
safety issues, it is common to provide proper shielding for workplaces to comply
with regulations, but less emphasis is placed on designing the work environment

in such a way that the control panel and patient monitoring devices are located in order to minimise staff distraction during irradiation. A quality management system, including a programme of quality control checks, is essential, but many double-checks may be rendered inefficient if the staff function 'mechanically' without due thought, or are subject to continual distractions.

2.3.1. Recommendation

(21) The establishment of a safety culture is of paramount importance in the prevention of accidental exposures in radiation therapy. Good practice is necessary but not sufficient. Detection and avoidance of errors require going beyond good practice, since even a well-designed system of controls and verification can suffer degradation with time if not monitored continuously. Radiation therapy should be performed with a full understanding of the process, due thought, mindfulness, alertness, and a clear sense of accountability. Hospital administrators and the heads of radiation therapy departments are responsible for cultivating these qualities and attitudes, and for encouraging excellence, particularly in matters related to safety. A working environment that facilitates concentration, avoids distraction, and promotes a questioning and learning attitude by the staff is essential for safe clinical operation.

2.4. Lessons from acceptance, commissioning, and calibration

(22) Beam calibrations are performed during commissioning or after a repair that may affect the beam characteristics. There are many opportunities for error in the determination of absorbed dose or dose rate, and these can lead to under- or over-dosage of all treatments on the miscalibrated machine. When the error is large enough to cause death or severe complications, such accidental exposures are clearly of a catastrophic nature. Potential sources of error include: mispositioning the ionisation chamber, misunderstanding or misreading the ion chamber calibration certificate, misreporting the irradiation parameters used in the calibration, temperature and atmospheric pressure corrections, errors with any of the series of correction coefficients, or simply an error in the calculation. Case histories of this type of accidental exposure are given in ICRP (2000) and IAEA (1998, 2000a). In addition, it is also possible that the absorbed dose is correctly determined but incorrectly entered into the TPS.

2.4.1. Recommendations

(23) The number of potential errors in dose calculations can be reduced significantly by using well-proven spreadsheets based on widely accepted protocols (IAEA, 1997, 2000a). However, measures should be taken to avoid misuse of the spreadsheets.

(24) Errors can be detected by an independent absorbed dose determination. If two different people come to nearly the same result, the probability of a major undetected error is very low because it would require both individuals to commit exactly

the same error, or a different error of the same magnitude and in the same direction. Ideally, both absorbed dose determinations should be independent (i.e. they should not influence each other to avoid repeating the same error). One example of independence is the use of a postal thermoluminiscence dosimetry (TLD) audit, provided that the commencement of patient treatment can wait until the TLD results are available. These safety measures are also applicable to new technologies. As indicated in ICRP *Publication 86* (ICRP, 2000), every radiotherapy centre should participate regularly in an external audit programme to verify the calibration of treatment units.

2.5. Commissioning of treatment planning systems

(25) Poor understanding of TPSs has led to severe accidental exposures involving large numbers of patients. Mistakes have included entering the wrong basic data upon which dose calculations rely, such as the absorbed dose at the reference point, depth doses, dose profiles, and wedge factors. Another reported error was applying correction factors for distance and wedges twice, as a result of not being aware that the TPS had already included these corrections in the calculation of the treatment dose or monitor units (MUs). For radioisotope units, accidental exposures have resulted from using the wrong decay rate (half-life) or the wrong date associated with the initial source activity or initial absorbed dose rate (IAEA, 2000b, 2001; ICRP, 2000).

2.5.1. Recommendations

(26) Formal purchasing agreements should include provisions to ensure that manufacturers provide appropriate training so that the staff are fully acquainted with the new system before it is introduced clinically. Systematic commissioning of a TPS is as important as the commissioning of the treatment machine. There are well-recognised international protocols that can be used as guidance (Fraass et al., 1998; IEC, 2000; ESTRO, 2004; IAEA, 2007c, 2008b) for the tests to be performed during acceptance testing and commissioning of a TPS. Operation of a treatment preparation or delivery system should never rely upon verbal transmission of procedures or operating instructions. Rather, instructions for use should be written, unambiguous, and in a language understandable to the users. All of these measures are applicable to new technologies.

2.6. Treatment-related lessons

2.6.1. Treatment preparation

(27) A major reported accidental exposure resulted from a change from normal procedure in the use of the TPS without validating the new procedure, and without an independent calculation of the MUs (IAEA, 2001).

(28) Nowadays, such calculation of MUs, designed to check the TPS calculation, is performed using an in-house-developed spreadsheet or a commercial MU calculator. Accidental exposures might occur due to the transfer of in-house-developed spreadsheets from one user to another without full understanding of the algorithm and user interface by the recipient.

(29) Errors have been made during treatment simulation due to the incorrect labelling of images; for example, a left–right swap, resulting in treating the wrong side of the patient. Such incorrect labelling could also appear in other stages of image handling during treatment preparation.

Recommendations

(30) Deviations from the manufacturer's instructions should be avoided. When this is not possible, proposed deviations should be discussed thoroughly with the manufacturer and be subject to specific tests and validation before use for clinical treatments. Calculation of the number of MUs for each patient independently from the TPS would have avoided most of the major accidental exposures resulting from the misuse of a TPS. Commissioning and regular quality control checks need to be applied to in-house-developed spreadsheets for MU calculations and any commercial MU calculators. This is even more important on receipt of an in-house-developed spreadsheet from another radiotherapy department or another user.

(31) These recommendations are also applicable to new technologies, although the independent calculation of MUs is not as simple as with conventional techniques and requires more complex MU independent verification software.

(32) In-vivo dosimetry can detect deviations in dose from that prescribed at the entrance or exit of the beam. These deviations may arise not only from the determination of absorbed dose at a reference point, but also from errors in the calculation of treatment doses and in treatment set-up. For some new technologies, such as IMRT, in-vivo dosimetry is more difficult to implement, especially if the same level of accuracy is expected as for conventional radiotherapy. In-vivo dosimetry specifically adapted to IMRT by means of diodes and metal oxide semiconductor field effect transistor (MOSFET) detectors have been reported, as have other promising new approaches such as transit dosimetry with electronic portal imaging devices used as detectors. This is discussed in more detail in Section 3.

2.6.2. Treatment delivery

(33) Errors during treatment set-up and delivery have been reported related to treating the wrong patient, treating the wrong anatomical site, or using the wrong dose or field sequencing. These errors may be as a result of using the wrong patient's chart without proper identification (i.e. by a photograph or similar method), confusing fiducial marks and tattoos, different patient positioning for simulation and treatment, wrong selection of parameters (e.g. setting the machine for rotational therapy rather than stationary treatment), failure to realise that the treatment of one of the sites was already completed, failure to introduce intended wedges, repeated or missing treatment fractions, and inadvertent couch movement during treatment.

Recommendations

(34) Identification of the patient, the treatment site, and the correct plan is essential at each step of the treatment process (including for patient-specific accessories). Provision for identifying the patient by a photograph is highly advisable, as is provision for an active response by patients to three unique identifiers (e.g. name, address, and date of birth), as well as identifying fiducial markers and tattoos. In addition, there is a need for identification procedures for patients who are unconscious, deaf, mute, or who do not speak the local language. Modern digital techniques make this approach simpler as every radiation therapy department can have digital cameras and the means to incorporate the picture into the treatment chart. Modern techniques can also be useful to further ensure identification in the future, such as individual identity cards incorporating bar codes or fingerprint identification. Some positioning errors can be automatically eliminated by patient data management systems which include R&V functions. Such systems require attention to other problems, such as relying too much on an automatic system as opposed to a manual system where the user is forced to maintain a higher degree of vigilance.

(35) Patient set-up errors can be avoided or detected through independent checks by two treatment delivery technologists. To maximise efficiency, it is necessary to identify those steps that are critical to safety. These are the steps that must be double-checked, and for which particular care must be taken in providing clear and appropriate instructions to the technologists.

2.7. References

Fraass, B., Doppke, K., Hunt, M., et al., 1998. American Association of Physicists in Medicine Radiation Therapy Committee Task Group 53: quality assurance for clinical radiotherapy treatment planning. Med. Phys. 25, 1773–1829.

ESTRO, 2004. Quality Assurance of Treatment Planning Systems. Practical Examples for Non-IMRT Photon Beams. European Society for Therapeutic Radiology and Oncology, Brussels.

IAEA, 1997. Absorbed Dose Determination in Photon and Electron Beams: an International Code of Practice. Technical Report Series No. 277, second ed. International Atomic Energy Agency, Vienna.

IAEA, 1998. Accidental Overexposure of Radiotherapy Patients in San José, Costa Rica. International Atomic Energy Agency, Vienna.

IAEA, 2000a. Absorbed Dose Determination in External Beam Radiotherapy: an International Code of Practice for Dosimetry Based on Standards of Absorbed Dose to Water. Technical Report Series 398. International Atomic Energy Agency, Vienna.

IAEA, 2000b. Lessons Learned from Accidental Exposure in Radiotherapy. Safety Report No. 17. International Atomic Energy Agency, Vienna.

IAEA, 2001. Investigation of an Accidental Exposure of Radiotherapy Patients in Panamá. International Atomic Energy Agency, Vienna.

IAEA, 2007a. On-site Visits to Radiotherapy Centres: Medical Physics Procedures. TECDOC-1543. International Atomic Energy Agency, Vienna.

IAEA, 2007b. Comprehensive Audits of Radiotherapy Practices: a Tool for Quality Improvement. Quality Assurance Team for Radiation Oncology (QUATRO). International Atomic Energy Agency, Vienna.

IAEA, 2007c. Specification and Acceptance Testing of Radiation Therapy Treatment Planning Systems. TECDOC-1540. International Atomic Energy Agency, Vienna.

IAEA, 2008a. Setting Up a Radiotherapy Programme: Clinical, Medical Physics, Radiation Protection and Safety Aspects. International Atomic Energy Agency, Vienna.

IAEA, 2008b. Commissioning of Radiotherapy Treatment Planning Systems: Testing for Typical External Beam Treatment Techniques. TECDOC-1583. International Atomic Energy Agency, Vienna.

ICRP, 2000. Prevention of accidental exposures to patients undergoing radiation therapy. ICRP Publication 86. Ann. ICRP 30 (3).

IEC, 2000. Medical Electrical Equipment: Requirements for the Safety of Treatment Planning Systems. IEC Publication 62083. International Electrotechnical Commission, Geneva.

3. SAFETY ISSUES WITH NEW TECHNOLOGIES

(36) This section provides an overview of the major new technologies and processes from the safety perspective, exploring them with the purpose of identifying potential weaknesses. As is often the case with predictive exercises, an initial degree of speculation may be unavoidable and, therefore, acceptable, provided that quantitative or semi-quantitative tools are subsequently used to assess likelihood of the scenarios and remove the unrealistic ones from futher study. In this context, this section is a preliminary exercise in safety assessment to familiarise oneself with the questioning approach. It is useful to compare potential weaknesses identified in this section with actual case histories of events described in Section 4 and Annex A, and with results of prospective safety assessments.

3.1. Justification issues

(37) The introduction of new technologies in radiation therapy is principally aimed at improving the treatment outcome by means of a dose distribution which conforms more strictly to the tumour (clinical target) volume. A highly conformal dose distribution allows for a dose escalation to the target volume without an increase in the radiation dose to normal tissues, or for a decrease in normal tissue dose without reducing the tumour dose, or a combination of both.

(38) There are a number of preliminary data confirming this expectation and suggesting a reduction in toxicity or an improvement in tumour control in terms of re-lapse-free survival. For example, the use of IMRT and tomotherapy has reduced xerostomia; dose escalation for prostate cancer in four randomised trials has improved relapse-free survival, which was not possible with conventional techniques; and global early and late toxicity have decreased dramatically from pelvic irradiation, mainly for prostate and cervical cancer, with modern radiation therapy (Cahlon et al., 2008; Kuban et al., 2008; Lee and Le, 2008). On the other hand, there are also studies which suggest that new technology has not yet resulted in a substantial improvement in the long-term outcome for most patients (Soares et al., 2005). There is clearly a need for systematic and precise evaluation of the benefits, as it would be unreasonable to use costly, time-consuming, and labour-intensive techniques for cases in which the same results could be obtained with conventional techniques which can be used with confidence and safety.

(39) If a decision is taken to use the new technologies, time dedication, training, and competence of staff need to be re-assessed. Once these issues have been addressed properly, a smooth, step-by-step, and safe transition over several years is necessary to maintain safety. Failure to do so may not only be a waste of resources but may also increase the likelihood of accidental exposures of patients.

3.2. Safety issues related to equipment design, acceptance testing, and commissioning

(40) Greater conformality of the dose distribution delivered to the patient may be obtained by:

- providing technical solutions to improve the conformity of the dose distribution to the target (e.g. intensity modulation, possibly combined with gantry rotation, stereotactic convergent multibeam approaches, and hadron therapy);
- providing treatment planning tools to optimise the dose distribution for each of these new technical solutions (e.g. inverse planning); and
- providing the means to apply them accurately to individual patients (e.g. image guidance and motion management).

(41) Most recent advances in radiation therapy have only been achievable through the increasing complexity of both equipment and treatment techniques. However, complexity may also increase the opportunities for accidental exposures. In order to obtain the desired improvements in clinical outcome without an increase in risk, appropriate safety provisions and strategies are necessary. The challenge is, therefore, to implement new technologies in conjunction with the appropriate means to ensure that they can and will be used safely.

(42) The most important feature related to the complexity and sophistication of 'new technologies' is the requirement for computer control. Computers are increasingly used at each stage of the process, from prescription to completion of the treatment. As a result of the complexity of many newer treatment strategies, 'common sense' and intuition may no longer be as effective a mechanism to perceive 'when something may be wrong' as it is with conventional radiation therapy (Rosenwald, 2002). In a conventional two-to-four field technique, it is feasible for someone with the appropriate knowledge and experience to identify a dosimetric error by looking at the treatment time or MUs, and making some simple calculations to confirm that these machine settings will result in the delivery of the correct dose to within a few percent. This reliance on common sense is no longer feasible in IMRT, for example, in which, instead of a collimator with four jaws and a relatively simple control mechanism, MLCs with 80 or more computer-controlled leaves are used to generate many elementary segments, applied either in a discrete sequence (step and shoot mode) or dynamically (sliding window).

(43) In risk reduction efforts, the role of the manufacturers is of increasing importance. As far as software is concerned, there is a need for the design of informative warnings, self-test capabilities, self-explanatory user interfaces, and internal safety interlocks to prevent improper use that may lead to accidental exposures. Technology- and technique-specific training become major issues for users, as well as installation and maintenance engineers.

(44) When introducing a new or upgraded piece of equipment or software in a radiation therapy department, the process involves planning, purchase, installation, and acceptance testing. Acceptance testing is the process whereby the new item is tested against predefined specifications agreed with the vendor and reflected in the purchase contract. Upon acceptance, the new item is declared compliant with the purchase order, which means that the vendor may be paid and that the warranty period may start. However, before the equipment or software is used clinically, it is essential to complete a commissioning phase; this requires significant time and effort. The commissioning phase includes the calibration, characterisation, and customisation

of the system for clinical implementation at the user's site. Acceptance and commissioning are the user's responsibility. In this context, subcontracted commissioning performed by the vendor or a third party company, without local staff taking an active part, increases the potential for local people to be unaware of important information about the equipment, including safety issues.

(45) Specific training on the particular equipment to be used is normally agreed during the planning and purchase processes. The commissioning phase is an opportunity to complete the staff training, and to ensure that all staff who will use the equipment or software are familiar with it and its associated safety features, and have developed the required expertise for safe and effective use. 'Emergency procedures', including simulation of systems failures (e.g. interruption of a dynamic wedge or IMRT treatment, replanning a patient on a different machine, etc.), are also part of the acceptance and commissioning phases. It is important to be aware that if a calibration or basic data input error occurs during the commissioning phase, it will potentially be propagated to all patients planned or treated with this hardware or software component. New technologies are not different in this respect from conventional technologies, but the increase in complexity entails new challenges when designing comprehensive commissioning and quality management systems. Risk-informed strategies to focus on vital aspects of commissioning and training are given in Sections 4 and 5.

(46) Each step of the treatment process entails specific risks. The major steps are:

- treatment prescription;
- treatment preparation; and
- treatment delivery.

(47) These steps are integrated into a workflow that makes extensive use of computer resources, and requires effective and efficient patient data management. However, patient data management systems are associated with their own risks, as discussed in Section 3.6.

3.3. Treatment prescription

(48) Treatment prescription is the responsibility of the radiation oncologist. It consists of choosing a therapeutic dose to be delivered to a specified target volume according to a specified time pattern or fractionation. The acceptable doses to organs at risk are also part of the prescription. Standardised nomenclature and definitions of the various components of the prescription are helpful to avoid ambiguity and misinterpretation of the prescription within a single institution, or when sharing experience among several institutions. This standardisation has been facilitated by ICRU, which has provided recommendations about volume definition and dose reporting for 'conventional' treatments (ICRU, 1993, 1999).

(49) The dose distribution in the target for an IMRT plan can be considerably less uniform than that delivered with conventional approaches. This makes conventional reporting of dose to a single specific point (the so-called 'ICRU reference point') no longer applicable. In the preface to ICRU Report 76 (ICRU, 2006), it is recognised

that 'in some modern irradiation techniques such as IMRT, the required accuracy relates not only to the dose level at one, or several, reference point(s) but also to the dose distribution'.

(50) Thus, IMRT and inverse planning require changes in the approach to dose prescribing, and prescriptions now need to be expressed in terms of dose–volume objectives (e.g. minimum and maximum doses within the target volume) and dose–volume constraints (i.e. maximum doses to specified volumes of the organs at risk).

(51) Clear recommendations at the national or international level and strict application of local protocols are indispensable to avoid inconsistency between treatments given within a single institution or in different institutions.

(52) An additional risk which can accompany more advanced techniques is related to dose escalation to the target volume, with the intention of achieving better tumour control probability while maintaining an acceptable probability of complications in normal tissue. This strategy requires an improvement in the dose conformality, achieved by a reduction of geometrical margins, accompanied by accurate and sophisticated imaging methods to confirm and monitor patient positioning. If this requirement for geometrical accuracy is not fully appreciated, there is a risk that dose escalation could lead to severe patient complications.

3.4. Treatment preparation

(53) Treatment preparation consists of all the tasks to be performed before the actual radiation delivery is initiated. Treatment preparation is sometimes called 'treatment planning' in a broad sense (Fraass et al., 1998a; IAEA, 2004) and includes the following steps:

1. patient immobilisation;
2. patient data acquisition, combined or not with virtual simulation[3];
3. image segmentation and structure delineation; and
4. beam definition and optimisation of the dose distribution (sometimes called 'dose planning').

(54) The workflow associated with these steps requires computer networking and data exchange between devices which may be from different manufacturers but which need to maintain full interconnectivity and interoperability. The risks related to data transfer will be dealt with in Section 5. The following text will address issues related to the equipment and procedures used at each individual step during the treatment preparation process.

---

[3] The conventional 'simulation' done on a simulator used to be a radiological verification of beam set-up. This task would normally fit between Steps 3 and 4. For three-dimensional (3D) conformal radiotherapy, conventional simulation tends to be replaced with 'virtual simulation', in which the beams are defined on a computer console (possibly in real time) on the basis of the 3D reconstruction of the patient anatomy. This is sometimes combined with Step 2 and partly covers Steps 3 and 4.

3.4.1. Patient immobilisation

(55) Complete immobilisation of the patient to maximise interfraction reproducibility and minimise intrafraction motion is the first step of the treatment preparation process. Immobilisation has always been challenging in radiation therapy. Modern immobilisation techniques do not present major hazards to the patient in terms of excessive temperature or chemical toxicity. The greatest risk might be the restriction of vital patient movements, such as normal breathing, vomiting, and movement to avoid collisions during gantry and couch movements. Effective mechanisms are required to enable the patient to warn the operator immediately if anything abnormal occurs, either during treatment preparation or beam delivery. One option is to give the patient access to a push button alarm. Such push buttons are used for patients treated with automatic breathing control where a valve inhibits patient respiration when the beam is on.

(56) The need for more accurate and precise patient positioning may encourage the use of more constraining patient immobilisation devices (e.g. masks with or without bite blocks, stereotactic frames, etc.), and more attention to the associated risks described above may be required. On the other hand, the development of IGRT may provide more flexibility thanks to the possibility of correcting small deviations from ideal alignment of the patient with respect to the treatment beams at the time of treatment.

3.4.2. Patient data acquisition and virtual simulation

(57) Most modern treatment plans are based on CT data. Increasingly, other imaging modalities are added to provide more accurate delineation of the target volumes and organs at risk. Virtual simulation, which follows patient data acquisition, is becoming more widely used. When appropriate, virtual simulation can be performed while the patient is still present, which allows skin marking of the simulation beam projections, or after the patient has left, which requires a co-ordinate system referenced to radio-opaque markers present during imaging.

Patient orientation for imaging

(58) With the increased use of CT images, identification of the patient and, importantly, patient left–right orientation becomes crucial. Although the most common patient orientation is 'head first–supine', it is sometimes necessary to use other orientations [e.g. having the patient prone to treat the spinal cord (medulloblastoma), or 'feet first' to treat a leg]. In any case, a clear distinction is needed between the orientation used for CT data acquisition, for treatment planning, and for treatment delivery. It is expected that the orientation would remain the same for all steps, and that the consistency would be guaranteed by the use of digital imaging and communication in medicine (DICOM) standardisation. However, there are many possible combinations and special circumstances where it might be necessary to use a certain orientation for one step of the process but to 'declare' another orientation for one or more other steps. There is then a potential for error that could lead to severe accidental exposures.

Tissue density values from CT data and image distortion from imaging modalities

(59) The relationship between CT numbers and tissue density is used for dose calculation, in particular for heterogeneity corrections. Errors can be made with this correction, such as selection of the wrong correction table. Also, if artifacts are present or if contrast medium is being used, this might be misinterpreted as patient tissue density on a CT scan. Magnetic resonance imaging (MRI) can produce image distortion. Combining several imaging modalities brings an additional risk of misregistration[4], resulting in a significant error in the location of either the target or a critical structure.

Consistency of co-ordinates and beam characteristics

(60) Virtual simulation used for 3D conformal radiotherapy requires that the co-ordinate origin and conventions are correct and consistent between image data acquisition, virtual simulation, dose planning, and treatment delivery. A potential error arises from the use of the beam characteristics defined at the time of simulation (sometimes called 'set-up fields') for treatment planning or delivery without checking that all beam parameters that could influence the number of MUs or the dose calculation are correctly included. For example, accessories which are not required for the simulation process, such as trays and wedges, need to be added for the treatment as necessary. It is also possible that the simulated beams may not actually be used for treatment or are modified in some way, but are mistakenly kept by the system and then used.

Exposure from image acquisition and treatment planning

(61) The growing importance of imaging in the treatment planning process and the reduction of CT image acquisition times are likely to increase the number of examinations with x-ray diagnostic energies. In addition, post-treatment examinations are also required. Therefore, patient exposure to x rays of diagnostic energies, both pre and post treatment, tends to increase. Exposure from localisation imaging during the course of treatment is discussed in Section 3.5.1. One example of the use of CT is the use of 4D acquisition where each single slice is replaced potentially by 10 slices acquired at different phases of the respiratory cycle. This additional exposure from imaging, which was considered negligible until recently, could increase (ranging up to 0.1 Gy) and needs to be assessed (Murphy et al., 2007).

3.4.3. Image segmentation and structure delineation

(62) Image segmentation is an image processing method to enhance and distinguish an object or its boundaries (lines, curves) from the rest of the image. In radiation therapy, this is typically used to locate and delineate organs or structures more clearly in images. This phase of treatment preparation is partly performed manually and partly performed using either TPS tools or software provided specifically for vir-

---

[4] Registration is the process of linking image data from different examinations to a single co-ordinate system; typically that of the treatment planning CT.

tual simulation. The radiation oncologist delineates the anatomical structure (target volume and organs at risk), but such delineation is often interpreted in different ways. There could be substantial deviations depending on the training and expertise of the clinician. The use of atlases validated by expert consensus could reduce this cause of uncertainty.

(63) There are also a number of potential errors associated with the software tools that are available and data handling in the subsequent treatment planning phase. For example, the system may be capable of automatic external or internal structure extraction or 3D expansions. In most cases, improper tools or the improper use of tools will result in a loss of geometric accuracy, with possible dose deviations at the periphery of the target volume and in normal tissues without changing the dose at the reference point. Dose planning algorithms may include constraints on how the anatomical model should be prepared (e.g. number or spacing of slices, contour regularity, absence of intersections or overlapping regions, CT number allocations), and these may not be explicitly identified by embedded warnings or interlocks. If such warnings and interlocks are not present and the user does not understand the limitations of anatomical modelling, severe errors may appear in the subsequent handling of these data for beam set-up or dose calculations.

(64) The following example illustrates this situation. Some algorithms do not reconstruct a continuous patient surface from a limited number of CT slices. When the beam axis crosses the space between slices, acquired at specific positions with considerable distance in between, or when a non-coplanar beam impinges on the upper or lower section that is considered by the anatomical model as 'empty', the calculated dose distribution and number of MUs would be subject to error. Another error may occur when the user forgets to assign a density to a slice defined from a contour.

3.4.4. Beam definition and optimisation of the dose distribution

(65) This step, which is carried out using a computerised TPS, essentially consists of calculating the dose distribution for the beam set-up proposed for treatment. However, TPSs are no longer restricted to the computation of dose distributions; they also transform the radiation oncologist's prescription, consisting of dose, volume, and time pattern, into an optimised plan, ready to be transferred to the treatment machine. Some past accidental exposures have been caused directly or indirectly by the use of TPSs (IAEA, 2001), and it is generally recognised that the main source of error comes from a poor understanding of some of the functionalities of the TPS, particularly if combined with a lack of independent determination of the number of MUs (absence of independent double-checks and/or absence of in-vivo dosimetry). Before using a TPS for clinical treatments, a complex commissioning phase is indispensable as the incorrect input of basic parameters may lead to systematic errors affecting many patients. In addition, sporadic errors may occur in daily use of the TPS.

(66) In the context of this section, it is useful to clarify that the concept of optimisation is also a basic principle of radiation protection which establishes that the dose

should be as low as reasonably achievable. When this basic principle of optimisation of protection is applied to radiotherapy patients, the meaning is that 'exposure of normal tissue be kept as low as reasonably achievable consistent with delivering the required dose to the planning target volume, and organ shielding be used when feasible and appropriate' (IAEA, 1996). Keeping in mind that the requirement for doses to be as low as reasonably achievable only refers to normal tissue, it becomes clear that radiation protection principles and requirements are perfectly compatible with, even similar to, the concept of optimisation of radiotherapy itself.

(67) A list of the typical main tasks to be performed when using a TPS is presented in Table 3.1. A risk index has been given for each task. The risk index combines an estimation of the probability of occurrence and the risk of significant clinical consequences (severity) of an error resulting in a geometric or dosimetric deviation from the 'correct' plan. The purpose of this table is to help build a safety system with emphasis on the most hazardous steps, such as the beam model, the use of wedges or other beam modifiers, the management of beam weights, and the computation of MUs. The risk indices are defined according to the following scale:

1. high probability, low severity;
2. low probability, high severity; and
3. high probability, high severity.

(68) Modern TPSs are very complex and offer a full range of functionalities with many possible pathways. This makes formal acceptance testing and commissioning procedures to check all modes of operation even more important. All systems are susceptible to failure when operated using unusual pathways or outside of usual ranges. Most of the failures would result in a system crash with no consequences other than a loss of data and time. However, in some very unusual circumstances, failures could influence the treatment outcome. Controlling the clinical impact of such circumstances by instituting preventive actions (e.g. through redundancy) is precisely the aim of a systematic safety assessment approach, as described in Section 5. Risks can be reduced by providing the system with a fluent user interface, and sufficient warnings and interlocks. However, the greatest risk is associated with human error related to inappropriate use of the functionalities of the system, as a result of insufficient training or inadequate understanding of some aspects of the TPS. Such errors have direct implications for the quality and safety of treatment. Very severe accidental exposures of this type have already occurred in conventional radiation therapy. A simple secondary MU calculation, independent from the TPS, has proven for many years to be an efficient tool for prevention of major errors in dose delivery. With more complex therapy (e.g. IMRT), manual calculation is no longer feasible; however, computer programs, independent from the TPS, meet the same objective.

3.5. Treatment delivery

(69) The actual delivery of treatment can start as soon as the plan calculated by the TPS has been received by the treatment machine. The detailed transfer process is

Table 3.1. List of tasks performed when using a treatment planning system (TPS).

Task	Risk index	Comment
Preparation of the beam data library (parameterisation)	3	Critical step, particularly regarding reference dose rate, output factors, and machine geometric characteristics such as origin and orientation of scales
Patient anatomical data acquisition and data transfer to TPS	2	Main risk is related to patient orientation management (see Section 4)
Delineating the external contour and building the patient anatomical model	1	This directly influences the tissue thickness used for monitor unit, MU, calculations; there is a specific risk from top and bottom slice characteristics, especially for non-coplanar beams
Definition of shapes and densities for inhomogeneous regions	1	This directly influences MU calculations
Target and critical organ delineation*	1	This directly influences beam set-up and dose–volume histograms
Target volume expansion	1	This directly influences beam set-up and dose–volume histograms
Choosing treatment machine, modality, and energy	2	If using obsolete data, there is a risk of lack of consistency with actual equipment characteristics
Beam set-up definition (only for three-dimensional conformal radiotherapy)	2	The critical issues are: the distinction between source to skin distance, SSD, and isocentric techniques, the meaning of the displayed co-ordinates (e.g. if SSD is different from source to axis distance, SAD), and the interpretation of collimator and table rotations scales
Defining field shape	1	No serious risk if safety features are embedded to prevent wrong input (cf. Panama accident, IAEA, 2001)
Adding beam-modifying devices (shielding blocks tray, wedge filters, compensators, etc.)	3	Critical here is the awareness of the presence and nature of the modifiers since they have a strong influence on MUs
Choosing beam weighting points	3	Critical in this step is the avoidance of a beam weighting point that is lying in a region where dose is low (e.g. under a block) or where the dose gradient is high (e.g. field edge)
Defining (total or fractional) beam weighting (contribution)	2	This directly influences MU calculations. It is sometimes difficult to understand the exact meaning
Dose distribution calculation and display	1	An awareness of the calculation options is critical
Dose–volume histograms calculation and display	1	The critical issue is an awareness of the calculation options and volume definitions (are structures completely enclosed in the sampling region?)
Decision on final approved treatment plan	2	When several studies have been performed, the critical issue is assurance that the approved plan will actually be used for treatment
MU calculation	3	Could have been achieved prior to approval of the final plan or could be performed on a separate system; critical step is verification of all relevant data
Data transfer from TPS to treatment machine	3	Another critical step where the verification of all relevant data is necessary

Adapted from: SFPM, 2010. Recommandations pour la mise en service et l'utilisation d'un système de planification de traitement en radiothérapie (TPS). Société Française de Physique Médicale, in press. Available at: http://www.sfpm.fr.
MU, monitor unit.

* In radiotherapy departments where delineation is done in an external system and then imported as a digital imaging and communication in medicine file, attention to errors in data transfer to the TPS is required.

part of patient data management and is discussed in Section 3.6. In order to ensure that the treatment is given accurately at the proper anatomical location, the patient and beam set-up have to be kept consistent with the plan prepared during virtual simulation and/or dose planning.

3.5.1. Verification of patient and beam set-up

(70) Verification of the patient's position relative to the beams requires complete immobilisation of the patient. The risks related to fixation and immobilising devices have been addressed in Section 3.4.1.

Co-ordinates and external marks and references

(71) Historically, patient set-up has relied on skin marks placed at the centre and/ or the edges of the treatment fields during simulation. In modern systems using virtual simulation, the skin marks are often made at the beginning of the scanning, and therefore a translation (shift) of the table is often needed to align the patient reference co-ordinate system with the co-ordinate system on the accelerator.

(72) The patient reference co-ordinate system is defined with respect to radio-opaque and/or tattooed skin markers. On set-up, these may be aligned with the light projection from the wall-mounted lasers, while the required shift is determined during virtual simulation or from dose planning. Shifting is necessary unless the patient is firmly fixed to the table with an indexing system allowing the use of absolute co-ordinates (i.e. the same for localisation, planning, and treatment delivery).

(73) Relative shifting is preferably carried out with table scales or by measuring with a ruler. Although some systems allow the table co-ordinate origin to be reset to the origin of the patient reference co-ordinate system, in most systems, the shift needs to be accounted for by an addition or subtraction from the actual absolute table position. It is then possible to commit an error and use the wrong value or wrong direction for patient positioning. This risk is higher for the new generation of treatment techniques (i.e. fixed-beam IMRT, tomotherapy, robotic multibeam treatments, or VMAT) where it is difficult to make the link between beam orientation and position, and patient anatomy. For such techniques, light fields are either not present (as in tomotherapy) or are useless, and extensive use of imaging for verification of patient set-up is required.

(74) As a consequence of this 'new' approach to working with geometrical co-ordinates, technologists may tend to focus on the co-ordinates and lose very basic awareness of what they are treating (e.g. treating the correct side in the case of a lateral tumour). There may also be a tendency to rely quite heavily on the R&V system, which would not necessarily catch such errors (i.e. table co-ordinates without human supervision or 'too loose' tolerances).

Position verification against patient internal structures

(75) X-ray imaging is required to verify patient position and correct field shape and size through the visualisation of internal structures relative to the periphery of the field. This was traditionally achieved with radiographic portal films, which have been

replaced more recently by electronic portal imaging devices (EPIDs). Typically, verification takes place on the first day of treatment or 1 day before. It may be repeated several times at the beginning of the treatment course and weekly thereafter. With greater demand for conformality of the dose distribution to the target and margin reduction, accurate patient set-up is essential; this can only be guaranteed if image-based verification is repeated more often. A daily check could be recommended, but then the additional radiation dose received by the patient may not be negligible. This is particularly the case when low-sensitivity systems are being used (e.g. a liquid-ionisation-chamber-based EPID), but even with the development of relatively high-sensitivity amorphous silicon detectors, ignoring the contribution of this dose could result in an overdosage equivalent to one fraction over a full treatment course. Several solutions exist to deal with this problem: compensation for the additional verification MUs at each fraction, overall compensation at the end of treatment, or adjustment of the prescription and integration of the image dose contribution into the patient's treatment plan.

(76) A further difficulty arises because the region that has to be imaged for verification purposes is generally not limited to the target volume, and could include sensitive structures. This is well known from the double-exposure techniques which were used in conjunction with MV radiographic portal films. It also happens when standard beam orientations (typically anteroposterior and lateral) are used to verify the position of a patient treated with oblique incidence (coplanar or non-coplanar beams), and for more advanced IGRT techniques where the recommended protocol could be a daily acquisition of cone beam CT, serial tomography (tomotherapy), or standard in-room kV x-ray image pairs (e.g. proton therapy and robotic systems). In all cases, the resulting dose should be assessed and taken into account.

Correction of patient position

(77) Correction of patient position has traditionally been applied by technologists under the overall responsibility of the radiation oncologist. The radiation oncologist's decision, based on portal images (e.g. move the patient 0.5 cm in the craniocaudal direction and 1 cm to the left), was communicated to the technologist orally or in writing. This approach is being replaced by image or structure matching, performed manually or automatically, where the patient shift is designed to superimpose the current on-treatment image on a reference image [e.g. digitally reconstructed radiograph (DRR)] obtained from virtual simulation or from the TPS. The correctness of the shift is verified (or not) by comparison of the image in the corrected position (implying additional radiation dose) with the reference image. Apart from the additional exposure as discussed above, there is a risk of relying too heavily on the reference image, which may be incorrect due to errors made during virtual simulation or at the treatment planning stage. Such errors, if they occur, could be difficult to detect and would likely be present from the beginning until the end of the treatment course.

3.5.2. IMRT and other advanced dynamic techniques

(78) In the last decade, there has been an impressive development of IMRT techniques. This has been made possible by the advent of computer-based inverse planning

accompanied by technical solutions to accurately control the shape of MLC fields as a function of delivered MUs. More recently, tomotherapy and robotic radiation therapy have offered additional degrees of freedom, with the former allowing 360° rotation of the x-ray source around the patient concurrently with continuous table motion, and the latter providing optimised directions of hundreds of minibeams emerging from an accelerator mounted on a robotic arm. Other new techniques include VMAT and hadron therapy, employing protons and other heavy charged particles.

(79) The introduction of new technologies has generally been kept well under control, in spite of the inevitable pressure to achieve improvements and the desire to obtain the benefits. Specific quality control tests and procedures have been developed and published by advanced groups involved in prerelease testing before commercialisation, and/or provided by the manufacturers as part of the purchased 'package'. An example of such a procedure is the fairly common practice of pretreatment patient-specific validation on a phantom (sometimes called 'hybrid plan') before any IMRT treatment[5]. Disseminating these new techniques over a larger number of centres or involving a larger number of patients needs to be achieved progressively, accompanied by adequate quality control tests (IAEA, 2008). The following factors may increase the risk of accidental exposure if appropriate safety barriers are not implemented:

- Multiplication of the number of parameters to be monitored during treatment makes them more difficult to control.
- The mechanical aspects of the robotic approach used to control the spatial position of the accelerator or the patient is a potential source of danger (collision or failure of the control systems).
- The development of 'segmented' irradiation techniques, in which coverage of large irradiation volumes is achieved by the superposition of many elementary dose distributions, leads to stricter requirements on instantaneous dose rates. Such techniques could be static (step and shoot mode) or dynamic (sliding window, tomotherapy, multiconvergent robot driven beams, etc.). One important safety issue is the ability to stop the beam fast enough to avoid significant overdosage in the case of failure of a critical component. This is particularly relevant for dynamic techniques (including scanning beam approaches), where severe overdosage can occur if the mechanical or electronic system used to scan the beam over the entire treatment volume fails.

---

[5] Such a practice is representative of what should be done when implementing a new complex procedure where there is concern for some unexpected event with possible consequences. However, after time and with experience, these procedures could be revised and perhaps simplified. Other approaches include specific quality control of the equipment and systematic in-vivo dose measurements using diodes or MOSFET detectors (Higgins et al., 2003; Engström et al., 2005; Marcié et al., 2005; Piermattei et al., 2007; Alaei et al., 2009), and transit dosimetry which is expected to be easier to use and to provide better accuracy in highly modulated beams (McDermott et al., 2007; van Elmpt et al., 2008; van Zijtveld et al., 2009). At this point in time, it is too early to predict which of these methods or combination of methods will become the most widely used and will prevail.

(80) However, as in complex activities with potential for major accidents (e.g. aeronautical and nuclear industries), the technology is mature. Provided that the equipment is developed with proper consideration of safety issues according to industrial standards (IEC, 1997, 1998, 2000, 2005), and with reinforcement of safety interlocks compared with more traditional treatments (e.g. redundancy), the implementation of such advanced technology should not carry significant additional risks.

Exposure outside the target

(81) The segmental nature of most IMRT techniques implies an increase in the total number of MUs required for a given dose to the target. Therefore, the dose at a distance from the irradiated volume due to collimator leakage or head scatter, although generally negligible in conventional radiation therapy, might become significant for IMRT techniques. The neutron dose at a distance from target volumes treated with hadron beams and passive techniques (scattering foils and mechanical modulators) is also relevant to protection and safety. The quantification of these contributions and their impact on the risk of second cancers is still under debate, and is the subject of a separate ICRP/ICRU report (in preparation).

Complex dose measurements in combined small beams

(82) Dose measurements are more difficult to perform when small static or dynamic beams are combined to create the required dose distribution (IMRT, multibeam radiosurgery, tomotherapy, etc.). The choice of an appropriate detector (size, energy and temporal response, calibration, etc.) and the experimental set-up for beam calibration are of utmost importance (Alfonso et al., 2008). Although users are ultimately responsible for commissioning such systems properly, which should preferably be done with the help of more experienced colleagues and/or in the framework of structured networks or user groups, manufacturers should alert users to the configuration and complexity of such devices and their implications for dose measurement.

Software control of accelerator output

(83) The output of computer-controlled accelerators is difficult to predict using physical principles since the response of the controlling or measuring devices (e.g. the monitor chamber) can be corrected electronically or by software look-up tables which are more or less accessible to users. For example, the monitor chamber response is usually the result of a computation which can include, or not, corrections related to the collimator opening or presence of wedges. As an illustration, the output of one particular type of accelerator with 'enhanced wedges' varies rapidly as a function of the position of the fixed jaw, not as a function of the equivalent field size, whereas for another type of accelerator with virtual wedges, which are based on the same general principle, the output is practically independent of the position of the jaws (see Section 4.2.1). The reasons for these differences are rather complex, and may be explained by electronic control of the dose rate or cumulative dose as a function of the moving jaw position (Leavitt et al., 1997, Liu et al., 1998, van Santvoort,

1998; Faddegon and Garde, 2006). Full information and alerts from the manufacturer are needed to avoid errors during calibration and commissioning.

3.6. Patient data management

3.6.1. Description of patient data management systems

(84) The use of computerised data management systems as the backbone of the organisation of a radiation therapy department makes it necessary to have a clear perception of the general workflow, and to interface the different components in such a way that data exchanges are safe and reliable. There were several steps in the historical development of patient data management systems.

R&V systems

(85) R&V systems consist of a database interfaced with the treatment machine, containing, for each patient, the prescribed machine parameters (e.g. gantry angle, field size, and number of MUs/beam) obtained from the TPS or the simulator. For each fraction and each beam, these values are checked automatically against the actual machine parameters that are set manually. In the case of a deviation larger than some predefined tolerance, a warning is displayed and an interlock prevents the treatment from starting. Additionally, for each fraction, the main machine parameters actually used for the treatment are recorded and can be reviewed.

(86) The functionalities of R&V systems are of major importance for the radiation therapy process. One of them is verification of the machine set-up, which improves overall safety but also introduces some risks needing attention, as discussed in Section 3.6.2. The second is the possibility to automatically complete a document which traces the details of patient treatment in the form of an electronic patient chart, as discussed in Section 3.3.

Radiation therapy information systems

(87) Modern systems are no longer strictly R&V systems, but have expanded into radiation therapy information systems (RTIS) that integrate (more or less effectively) many different components of the patient workflow, such as management of administrative data (e.g. billing), management of internal or external resources (e.g. people, equipment, rooms), and scheduling, creation, and updating of patient treatment charts. These software systems may also include image management, with picture archiving and communications system functionalities. They may be huge and difficult to understand properly (Fraass, 2008). These systems are meant to interface with existing in-house databases and require assistance from computer specialists.

Direct control of machine parameters

(88) In addition to RTISs, with the development of IMRT, there are new requirements for direct software control of the machine parameters in order to automatically control the beam sequence and drive the components used for beam modulation (MLC, cumulative MUs, gantry rotation, etc.). Therefore, the system

is no longer limited to verifying that the machine parameters are correct, but assumes control directly. Other means need to be found to guarantee consistency between prescription and delivery.

Communication between different components

(89) Communication between the different components of an integrated RTIS is generally achieved by a standardised approach, such as a 'transmission control protocol/ internet protocol' network, DICOM format and DICOM radiation therapy based exchanges. This provides flexibility and allows integration of equipment from different companies. However, the variety and complexity of the available solutions can lead to many potential pitfalls, essentially due to the introduction of data in the optional fields of the DICOM format, or to the particular use of some planning or treatment data by one manufacturer's device that can be misinterpreted by other devices.

3.6.2. Machine set-up verification functionalities of R&V systems

(90) The verification functions of R&V systems were designed to increase the reliability and safety of the radiation therapy process against human error in daily treatment delivery. It has been demonstrated that R&V systems are effective in detecting random errors which are operator dependent (e.g. checking the presence of wedge filters or the number of MUs) (Macklis et al., 1998). However, they can also introduce new types of error (Fraass et al., 1998b; Patton et al., 2003), for the following reasons:

- The daily use of an R&V system has an impact on the state of mind of technologists who know that there is an automatic safety system working in the background; in spite of their professional sense of responsibility, they could tend to relax their attention compared with a manual system, which would be fully under their control. A typical example is the application of the treatment parameters for the wrong patient (Patton et al., 2003; Huang et al., 2005) by simply clicking on the wrong line as part of a process which is highly repetitive.
- The data of an R&V system are normally exported from the TPS through a network that should provide an error-free solution, provided that the whole system has been commissioned properly. However, the transfer can be incomplete for some treatments, which then require additional data to be input manually. Such manual input is prone to error, but with the false confidence of working with an 'error-free' system.
- As for TPSs, the number of possible pathways of so many functionalities is very large. Occasional errors may occur in certain circumstances, such as purposely modifying a plan for a treatment that has already started, or transferring a patient from one machine to another.
- R&V systems are strongly interdependent on the other components of the radiation therapy network. Some manufacturers are even integrating the TPS functionalities as one module of the full RTIS. Thus, although a number of safety interlocks are normally included, it has become very difficult to understand the possible consequences of any action performed on the patient electronic chart. This may lead to increased risk of misinterpretation and error.

- The use of an R&V system is likely to change many of the errors that would have been random for a manual-based treatment set-up into systematic errors (and therefore much more severe) (Fraass et al., 1998b; Huang et al., 2005). As an example, a one-time human error resulting in an incorrect field size setting or an unintended omission of a wedge would become systematic for the whole course of treatment if a human error is made when introducing the data into the R&V system. This consideration should not be interpreted as a reason not to use R&V systems, which have the advantage of giving access to statistics on error rates, but as a warning about the potential shift from one-time errors to systematic errors (Goldwein et al., 2003).

(91) Many of these errors can be circumvented by careful screening of the data stored in the R&V system by an authorised physicist entitled to formally and manually 'approve' the plan before treatment delivery. Such approval may be made mandatory at the beginning of treatment and following any alteration to the plan. However, it is difficult to find the proper reference against which this screening should be done. One option could be the hard copy of the TPS output, but the manual approval of an electronic chart by a physicist does not prevent all errors and thus there is a need to develop independent automated verification systems.

3.6.3. Electronic chart functionalities of R&V systems

(92) Automatic recording of treatment parameters for each fraction presents the possibility of dispensing with the traditional paper chart where technologists manually record the basic information related to each fraction (e.g. the number of MUs), for each patient, and each beam. A 'paperless' department thus becomes a possibility.

(93) However, such a change is complex and requires careful analysis for the following reasons:

- Availability and reliability of computer systems and data storage facilities deserves special attention. Back-up solutions require careful thought in advance to cover any failure situation.
- The patient chart is the traditional location for handwritten information relating to the treatment prescription, the treatment execution, and instructions for any change occurring during the treatment course. In principle, the replacement of a hard copy by a computerised system has the advantage of automatically ensuring that all categories of staff will fill in the necessary information. Therefore, it can be well adapted to the standard processes as defined in existing protocols. However, for certain situations, such as a modification after treatment has started, efficient and safe communication between professionals becomes crucial. This will depend on the tools available in the computerised system and on the local rules for describing unexpected events that happen during a treatment course.
- Special attention needs to be paid to the follow-up from the delivery of each fraction. Most R&V systems offer the possibility of accumulating the dose at one (or several) reference point(s) as the treatment proceeds, and giving a warning if the

dose (or number of fractions) exceeds the prescribed value. There are several possibilities for misreporting in the electronic chart, and these should be investigated carefully. Examples include machine failures, treating the patient on another machine, or treating the patient on a non-working day. If technologists rely too heavily on automatic alerts by the system, in cases where they do not work or do not apply, there is the risk of either repeating a session that has been delivered but not been recorded, or stopping treatment before the end of the course.

(94) It is difficult to review the numerous possible pitfalls related to the introduction of an electronic chart into a radiation therapy department. It is, therefore, important to develop thorough procedures and to plan a commissioning phase and a 'probing period' to ensure that such a system is used safely. Prospective systematic approaches to explore what could go wrong are described in Chapter 5, and they can be used as tools to develop such procedures and commissioning plans.

3.6.4. Image handling

(95) Images are becoming an essential component of the patient record, and some systems incorporate them directly into the RTIS. These images may be:

- pictures taken to confirm the patient's identity;
- pictures taken to help in patient set-up;
- diagnostic images;
- images used in patient anatomical reconstruction;
- reference images, either acquired directly at a simulator or digitally reconstructed from volumetric data (i.e. DRRs);
- portal images used to confirm the field shape and its position with respect to anatomical structures; and
- verification images (kV or MV images of orthogonal verification beams, CT or cone beam CT images) acquired in the treatment room to confirm patient position with respect to the isocentre of the machine.

(96) In all cases, careful image identification is required and there is a need to know, without ambiguity, the patient to whom the images belong, the date and time, the device used for imaging, the geometrical imaging characteristics, the doctor or technologist who performed the image acquisition, and the related beam and session number whenever relevant. Image (left–right) orientation with respect to the patient and the co-ordinate system of the treatment unit must also be known precisely. DICOM standardisation helps to ensure that such information is attached to digital images, and automatic image transfer reduces the risk of error. However, in many cases where image-related information needs to be completed manually, errors in this action could propagate throughout the whole chain.

(97) The most frequent and most severe cause of error may be related to the reference images used to check or adjust the beam position with respect to the patient. If these images are manually attached to the relevant beam in the database, there is a possibility that another beam of the same patient or a similar beam from a plan other

than the approved plan could be used as the reference. The consequence could be a systematic geometric mismatch that would remain undetected throughout the treatment course. In all cases, images play an important role in the safety and precision of treatment delivery, but errors could occur, leading to significant consequences for treatment outcome.

3.7. References

Alaei, P., Higgins, P.D., Gerbi, B.J., 2009. In vivo diode dosimetry for IMRT treatments generated by Pinnacle treatment planning system. Med. Dosim. 34, 26–29.

Alfonso, R., Andreo, P., Capote, R., et al., 2008. A new formalism for reference dosimetry of small and nonstandard fields. Med. Phys. 35, 5179–5186.

Cahlon, O., Hunt, M., Zelefsky, M.J., 2008. Intensity-modulated radiation therapy: supportive data for prostate cancer. Semin. Radiat. Oncol. 18, 48–57.

Engström, P.E., Haraldsson, P., Landberg, T., Hansen, H.S., Engelholm, S.A., Nyström, H., 2005. In vivo dose verification of IMRT treated head and neck cancer patients. Acta Oncol. 44, 572–578.

Faddegon, B.A., Garde, E., 2006. A pulse-rate dependence of dose per monitor unit and its significant effect on wedge-shaped fields delivered with variable dose rate and a moving jaw. Med. Phys. 33, 3063–3065.

Fraass, B.A., 2008. QA issues for computer-controlled treatment delivery: this is not your old R/V system any more! Int. J. Radiat. Oncol. Biol. Phys. 71 (Suppl.), S98–S102.

Fraass, B., Doppke, K., Hunt, M., et al., 1998a. American Association of Physicists in Medicine Radiation Therapy Committee Task Group 53: quality assurance for clinical radiotherapy treatment planning. Med. Phys. 25, 1773–1829.

Fraass, B.A., Lash, K.L., Matrone, G.M., et al., 1998b. The impact of treatment complexity and computer-control delivery technology on treatment delivery errors. Int. J. Radiat. Oncol. Biol. Phys. 42, 651–659.

Goldwein, J.W., Podmaniczky, K.C., Macklis, R.M., 2003. Radiotherapeutic errors and computerized record/verify systems. Int. J. Radiat. Oncol. Biol. Phys. 57, 1509–10.

Higgins, P.D., Alaei, P., Gerbi, B.J., Dusenbery, K.E., 2003. In vivo diode dosimetry for routine quality assurance in IMRT. Med. Phys. 30, 3118–3123.

Huang, G., Medlam, G., Lee, J., et al., 2005. Error in the delivery of radiation therapy: results of a quality assurance review. Int. J. Radiat. Oncol. Biol. Phys. 61, 1590–1595.

IAEA, 1996. International Basic Safety Standards for Protection against Ionizing Radiation and for the Safety of Radiation Sources. Safety series No. 115. Food and Agriculture Organization of the United Nations, International Atomic Energy Agency, International Labour Organization, OECD Nuclear Energy Agency, Pan American Health Organization, World Health Organization, International Atomic Energy Agency, Vienna.

IAEA, 2001. Investigation of an Accidental Exposure of Radiotherapy Patients in Panamá. International Atomic Energy Agency, Vienna.

IAEA, 2004. Commissioning and Quality Assurance of Computerized Planning Systems for Radiation Therapy of Cancer. IAEA TRS-430. International Atomic Energy Agency, Vienna.

IAEA, 2008. Design and Implementation of a Radiotherapy Programme: Clinical, Medical Physics, Radiation Protection and Safety Aspects. International Atomic Energy Agency, Vienna.

ICRU, 1993. Prescribing, Recording and Reporting Photon Beam Therapy. Report No. 50. International Commission on Radiation Units and Measurements, Bethesda, MD.

ICRU, 1999. Prescribing, Recording and Reporting Photon Beam Therapy (Supplement to ICRU Report No. 50). Report No. 62. International Commission on Radiation Units and Measurements, Bethesda, MD.

ICRU, 2006. Measurement Quality Assurance for Ionizing Radiation Dosimetry. Report No. 76, vol. 6, No. 2. International Commission on Radiation Units and Measurements, Bethesda, MD.

IEC, 1997. General Requirements for Safety. 4. Collateral Standard: Programmable Electrical Medical Systems. IEC 60601-1-4. International Electrotechnical Commission, Geneva.

IEC, 1998. Part 2: Particular Requirements for the Safety of Electron Accelerators in the Range of 1 to 50 Mev. IEC-60601-2-1. International Electrotechnical Commission, Geneva.

IEC, 2000. Medical Electrical Equipment: Requirements for the Safety of Treatment Planning Systems. IEC-62C/62083. International Electrotechnical Commission, Geneva.

IEC, 2005. Medical Electrical Equipment: Safety of Radiotherapy Record and Verify Systems. IEC-62C/62274. International Electrotechnical Commission, Geneva.

Kuban, D.A., Tucker, S.L., Dong, L., et al., 2008. Long-term results of the M.D. Anderson randomized dose-escalation trial for prostate cancer. Int. J. Radiat. Oncol. Biol. Phys., 71, 1288; author reply 1288–1289.

Leavitt, D.D., Huntzinger, C., Etmektzoglou, T., 1997. Dynamic collimator and dose rate control: enabling technology for enhanced dynamic wedge. Med. Dosim. 22, 167–170.

Lee, N.Y., Le, Q.T., 2008. New developments in radiation therapy for head and neck cancer: intensity-modulated radiation therapy and hypoxia targeting. Semin. Oncol. 35, 236–250.

Liu, C., Li, Z., Palta, J.R., 1998. Characterizing output for the Varian enhanced dynamic wedge field. Med. Phys. 25, 64–67.

Macklis, R.M., Meier, T., Weinhous, M.S., 1998. Error rates in clinical radiotherapy. J. Clin. Oncol. 16, 551–556.

Marcié, S., Charpiot, E., Bensadoun, R-J., 2005. In vivo measurements with MOSFET detectors in oropharynx and nasopharynx intensity-modulated radiation therapy. Int. J. Radiat. Oncol. Biol. Phys. 61, 1603–1606.

McDermott, L.N., Wendling, M., Sonke, J-J., van Herk, M., Mijnheer, B.J., 2007. Replacing pretreatment verification with in vivo EPID dosimetry for prostate IMRT. Int. J. Radiat. Oncol. Biol. Phys. 67, 1568–1577.

Murphy, M.J., Balter, J., Balter, S., 2007. The management of imaging dose during image-guided radiotherapy: report of the AAPM Task Group 75. Med. Phys. 34, 4041–4063.

Patton, G.A., Gaffney, D.K., Moeller, J.H., 2003. Facilitation of radiotherapeutic error by computerized record and verify systems. Int. J. Radiat. Oncol. Biol. Phys. 56, 50–57.

Piermattei, A., Cilla, S., D'Onofrio, G., et al., 2007. Large discrepancies between planned and actually delivered dose in IMRT of head and neck cancer. A case report. Tumori 93, 319–322.

Rosenwald, J.C., 2002. Safety in radiotherapy: control of software and informatics systems. Cancer Radiother. 6 (Suppl. 1), 180s–189s (in French).

SFPM, 2010. Recommandations pour la mise en service et l'utilisation d'un système de planification de traitement en radiothérapie (TPS). Société Française de Physique Médicale, in press. Available at: http://www.sfpm.fr.

Soares, H.P., Kumar, A., Daniels, S., et al., 2005. Evaluation of new treatments in radiation oncology: are they better than standard treatments? JAMA 293, 970–978.

van Elmpt, W., McDermott, L., Nijsten, S., Wendling, M., Lambin, P., Mijnheer, B., 2008. A literature review of electronic portal imaging for radiotherapy dosimetry. Radiother. Oncol. 88, 289–309.

van Santvoort, J., 1998. Dosimetric evaluation of the Siemens virtual wedge. Phys. Med. Biol. 43, 2651–2663.

van Zijtveld, M., Dirkx, M., Breuers, M., de Boer, H., Heijmen, B., 2009. Portal dose image prediction for in vivo treatment verification completely based on EPID measurements. Med. Phys. 36, 946–952.

4. REPORTED ACCIDENTAL EXPOSURES WITH NEW TECHNOLOGIES

(98) This section contains representative case histories of accidental exposures obtained from various sources, such as ROSIS, the event notification reports of the US Nuclear Regulatory Commission, and reports on ad-hoc investigations of specific accidental exposures. Each case relates to some specific new radiation therapy technology or procedure. These cases include some with very severe consequences. For each case, the summary of the case history is followed by a discussion on the lessons to learn from the event. Additionally, short descriptions of incidents without severe consequences, taken from ROSIS, are included in Annex A.

4.1. Events related to beam output and calibration

4.1.1. Calibration problems of small fields

Case 1. Inappropriate detector size used when commissioning micro-MLCs (ASN, 2007; Derreumaux et al., 2008)

(99) Micro-MLCs are able to form very small irradiation fields with high precision. These fields can be used when irradiating small targets in, for example, the brain, such as in radiosurgery applications. When commissioning a treatment unit equipped with a micro-MLC, beam data have to be collected using dosimeters with an appropriate detector size considering the potentially small size of the irradiation fields.

(100) In April 2006, a hospital physicist commissioned a new stereotactic unit capable of operating with micro-MLCs (3-mm leaf width at isocentre) or conical standard collimators. With this unit, it is possible to shape clinically usable fields down to the very small field size of 6 mm × 6 mm. When collecting beam data for the TPS, it is necessary to measure the beam dose characteristics down to this field size. The beam data are subsequently used for treatment planning purposes. Data collected for micro-MLCs are handled separately from data collected for standard collimators.

(101) When measuring absorbed doses and collecting beam data (scatter factors) for very small beams formed by the micro-MLCs, the physicist at the hospital used a Farmer $0.6 \, \text{cm}^3$ ionisation chamber, which is too large for this type of measurement. Consequently, the dose measurements were incorrect for all small micro-MLC fields, resulting in the calibration files for all micro-MLC fields being wrong. A maximum overdose of approximately 200% was administered as a result of this error when these fields were used. Patients treated with standard collimators were not affected.

(102) The anomaly in the calibration files at this hospital was discovered by the vendor some time later during a review of calibration files collected from several European centres. The vendor informed the hospital of the anomaly in April 2007. By this time, 172 patients had been treated stereotactically on the unit. One hundred and forty-five of these patients had been treated using micro-MLCs, and had thus been affected by the erroneous measurement. In most cases, the dosimetric

impact was assessed as having been small. However, tolerance doses in normal tissues and organs were exceeded in some patients.

Discussion and lessons. (103) Micro-MLCs pose new challenges to the physics community through the additional knowledge and expertise required for their use. Their commissioning and use, therefore, require additional staff training and, in particular, verification that physicists in the department have a thorough understanding of the new technology, its features, and the measurements to be performed. Specifically, full knowledge is required of the physics of the small fields produced with micro-MLCs, the conditions they impose on detector size, the dosimetric effects of partial irradiation of a detector that is larger than the beam cross-section, and the limitations of the protocols designed for measurements in larger beam sizes where electronic equilibrium conditions are met. Such conditions may not be present in small fields.

(104) After revisiting the training issues, preparation for commissioning is necessary. This includes the preparation of procedures for measurements with micro-MLCs, or a conscious decision to adopt relevant procedures from recognised protocols.

(105) Finally, independent checks of the measurements and calculations, followed by clarification of any discrepancy, are necessary before the radiation therapy equipment is used clinically. Independent checks would be strengthened by inviting a physicist from another hospital to confirm the measurements and calculations using their own equipment and calculation methods. With these measures in place, such an accidental exposure would have been very unlikely to occur.

4.1.2. Intra-operative radiation therapy beam calibration issues

Case 2. Intra-operative calibration error from the wrong calibration file (ROSIS, 2008)

(106) New intra-operative radiation therapy (IORT) equipment was delivered to a clinic. The clinic received no information from the manufacturer regarding how the absorbed dose was measured at specific distances from the intra-operative applicators, including measurement geometry, and thus how the pre-installed calibration files containing the information required for the calculation of treatment times were devised.

(107) A phantom was created at the clinic to measure the absorbed dose at the commissioning of the IORT equipment. During commissioning, it was noted that the two applicators with a diameter of 4 cm were equal in all geometrical aspects, but that the manufacturer had supplied the two applicators with calibration files (irradiation time required for certain dose) which differed from each other by 20%. Thus, when calculating the treatment times for giving a certain dose to a patient, the time was 20% longer for one of the two identical applicators. The local physicists, after verifying the numbers stated in the calibration files by phantom measurements, mentioned the anomaly to the company engineer who had installed the system. The engineer was of the view that the local physicists had not measured the absorbed

dose correctly for verification of the calibration files using the locally created phantom.

(108) Some time later, in connection with a meeting on other technical issues, the matter was brought up again with the company. The company realised that they had provided an incorrect calibration file for one of the 4-cm applicators, causing the dose to differ by +20% from the intended dose. Due to the low energy of radiation (50 kV), only 1.5–2 mm of extra tissue was irradiated with an excessive dose.

Discussion and lessons. (109) The most important point from this event is that when discrepancies in dose measurements are found, it is the ultimate responsibility of the hospital to investigate the situation thoroughly before applying the beam to patients. The reliance of the physicists on the informal opinion of an engineer suggests additional lessons for the hospital (i.e. the need to include a list of acceptance tests to be performed, and procedures to resolve discrepancies found during acceptance, commissioning, and afterwards in the purchase contract).

(110) However, it is also the responsibility of manufacturers, suppliers, and installers to deliver the correct equipment with the correct calibration files and accompanying documents, including measurement geometry used when creating calibration files. Effective internal quality control procedures are needed to identify errors before the equipment is handed over by the installers, A commitment by the supplier to provide correct information and advice following questions formulated by the hospital staff is also crucial. A lesson that suppliers could learn from this event is the need to ensure that training for their engineers addresses the tests to be performed and documented before and during acceptance. This training should include the advice to be given to hospital staff.

4.1.3. Beam output drift in tomotherapy

Case 3. Incorrect tolerance for the interlocks of a tomotherapy machine (Saint-Luc Hospital, Belgium, 2008)

(111) On a tomotherapy machine on which daily morning checks were performed systematically to assess beam output stability, a sudden drift was observed one morning, with an underdosage larger than 10%. The internal safety interlocks of the machine were not 'seeing' this drift. Further treatments of patients were cancelled, although there was doubt regarding the validity of the response of the local dosimeters used for these measurements that seemed to differ from the response of a dosimeter supplied by the manufacturer.

(112) After further investigation, it appeared that the difference between the readings of the local dosimeters and the manufacturer's dosimeter was only 1.3%. However, it was noticed that the safety threshold used for the interlock of the output of the tomotherapy machine had previously been set to a tolerance larger than ±10%. Examination of the log-book of treated patients revealed that three patients treated in the afternoon of the day preceding the faulty morning check were underdosed by 12%. It was later discovered that the incorrect setting of the interlock had been in

place since the original installation of the machine. Apparently, the installation technician enlarged the tolerance to facilitate adjustment of the beam output, and forgot to reset it to the correct value. It is still unclear why the interlock was set with such a wide tolerance and why there was a sudden drift of the machine output. Regardless, the magnetron and the target were replaced after the problem was discovered, and before the machine was cleared for clinical use.

Discussion and lessons. (113) The output verification of a radiation therapy machine is performed according to a predefined schedule, typically daily. The dose delivery for all patients treated between two consecutive checks relies on the stability of the machine and its dose monitoring system. In principle, internal safety interlocks should prevent incorrect output, but the tolerance to which they are set is not usually accessible to the user or checked as part of the local department quality control procedures. If the tolerance is too large and a problem occurs with the machine output between two consecutive checks, it could remain undetected until the next check. For a tomotherapy machine, the dose rate is critical because it is used in combination with the table translation to control the delivered dose. However, similar problems can be found on a conventional linear accelerator (e.g. with respect to the beam uniformity or to monitor response control).

(114) The lesson from this incident is that users need to understand how the beam is monitored and which interlocks are provided by the manufacturer. Interlock checks may need to be linked to acceptance tests. With the assistance of the manufacturers, the users should include a method in their departmental quality control procedure to check that the safety interlocks are set correctly, especially after maintenance or repair. In addition, manufacturers should develop more advanced solutions, preferably automated, to avoid machine parameters being set outside of the allowed range.

4.2. Events related to treatment preparation

4.2.1. Problems with dynamic wedges

Case 4. MU calculation for the wrong type of wedge (ASN, 2007; IGAS, 2007; Ash, 2007; Derreumaux et al., 2008)

(115) When a new treatment technique was being introduced at a hospital in 2004, it was decided to change from static mechanical (hard) wedges to dynamic (soft) wedges for the treatment of prostate cancer patients. When treating with open fields or using hard wedges in this centre, the standard practice was to verify MUs independently through calculation checks, as well as using diodes for an independent check of the dose delivered. The physicist involved in the change of technique was the only physicist working at the facility at the time, and was also on-call in another facility.

(116) As part of the introduction of the new treatment technique, two of the dosimetrists (TPS operators) were given two brief demonstrations on how to use

the software. When changing the treatment technique, the previous safety provisions given by independent calculation of MUs and verification with diodes were removed. The reasons for these practice changes were that the independent calculation software could not handle dynamic wedges, and the interpretation of diode results would have been much more difficult when using soft wedges compared with hard wedges.

(117) The terminology of the treatment planning software, including displays, was in English, as were the operator manuals. The operators were French. Some of the French operators misunderstood the abbreviated English display on the interface to the TPS, and mistakenly selected hard wedges (identified by angle) when intending to select the plan with soft wedges. The correct box to tick in the TPS software for soft wedges was indicated 'EW' (enhanced dynamic wedge) without any angle indication; this was not easily understandable and did not correspond with the terminology in French. When the treatment planning of a patient had been finalised and the dose distribution had been optimised for hard wedges, the parameters, including the MUs, were manually transferred to the treatment accelerator, and the dynamic wedge option was manually selected. The number of MUs calculated for hard wedges from the plan was much greater than the number of MUs that was needed to deliver the same absorbed dose with soft wedges. Consequently, the patients affected by this error received a higher absorbed dose than intended. The reason for the higher number of MUs is illustrated in Fig. 4.1 and Table 4.1.

(118) Between May 2004 and August 2005, at least 23 patients received an overdose of 20–35%. Between September 2005 and September 2006, four patients died as a result of this accident. At least 10 patients showed severe radiation complications, with symptoms such as intense pain, discharges, and fistulas. Regional authorities were informed the month following the accident, but national authorities only received information a full year after the accident occurred.

Discussion and lessons. (119) Two brief demonstrations to two dosimetrists is insufficient when changing from static wedges to dynamic wedges, as use of the latter is more complex and is accompanied by critical safety issues. More thorough and effective training is required. The difference in the number of MUs required when using static and dynamic wedges was not fully appreciated. This insufficient understanding was aggravated by the removal of provisions such as independent calculations of MUs and dose checks with diodes. In summary, insufficient understanding of a new, more complex technique, instructions and displays in a language not understood by operators, and removal of checks made the accident more likely to happen.

(120) One important lesson is that there may be a temptation to remove existing checks when they cannot be easily applied to a new technology. This is a challenge inherent in more complex technology. The appropriate safety philosophy is to increase supervision during the implementation of a new technique, and hence maintain the required level of safety. This approach should be taken even if it entails the design of new verifications or adaptation of the old methods. The decision should never compromise safety.

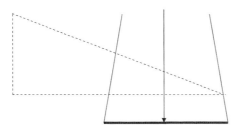

Enhanced dynamic (soft) wedge in this particular accelerator is
equivalent to a shifted physical wedge (variable attenuation at beam
axis depending on field size)

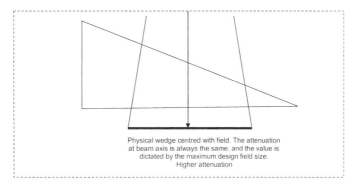

Physical wedge centred with field. The attenuation
at beam axis is always the same, and the value is
dictated by the maximum design field size.
Higher attenuation

Fig. 4.1. In this particular accelerator, when the dynamic wedge function is used, the collimator is almost closed at the beginning of irradiation and the jaws are aligned on the field edge where the higher dose is to be delivered (right side of the figure). As the irradiation begins, the moving jaw starts moving towards the opposed edge of the field (left side of the figure). This is analogous to a hard wedge with its thin edge aligned to the right field edge. A physical wedge in a symmetrical field, however, is centred with the beam (see lower part of the figure) and is designed to cover the largest field size. Thus, the thin edge of the physical wedge does not coincide with the field edge, except when the largest field size is selected. Therefore, for this particular accelerator design, it is only for the largest field size that the attenuation of the physical wedge would be equivalent to the dynamic wedge, and the ratio of monitor units (MUs) between the hard and soft wedges would be close to unity. For all smaller field sizes, the attenuation of the physical wedge is higher than the dynamic wedge, and the MU ratio is >1. The smaller the field size, the higher the MU ratio, as shown in Table 4.1.

4.2.2. Computer problems with IMRT

Case 5. Computer 'crashes' and loss of data in IMRT planning (VMS, 2005; NYC Department of Health and Mental Hygiene, 2005)

(121) A patient with head and neck cancer (oropharynx) was scheduled for IMRT at a radiation therapy facility in March 2005. An IMRT plan was prepared as per the standard institution protocol. A 'verification plan' (sometimes called a 'hybrid plan') was also prepared and tested prior to treatment, as required by the quality control procedures of the hospital, in order to verify that the calculated dose distribution would be achieved at irradiation. This verification plan confirmed the dosimetric correctness of the plan through use of portal dosimetry. Subsequently, treatment of the patient was delivered correctly for the first four fractions.

Table 4.1. Ratio of the numbers of monitor units (MUs) for physical/dynamic wedges for the same dose delivered at 15-cm depth using the isocentric technique and physical wedges of 45°.

Field size	Ratio of MUs for physical/dynamic wedges	
	6 MV	20 MV
5 cm × 5 cm	1.83	1.83
10 cm × 10 cm	1.55	1.61
15 cm × 15 cm	1.32	1.43
20 cm × 20 cm	1.12	1.27

Adapted from Rosenwald, J.C., 2007. Personal communication.

(122) After the first four fractions, a physician reviewed the case and concluded that there was a need to modify the dose distribution in order to reduce the dose to specific organs at risk. This task was given to a dosimetrist who started by copying the treatment plan in order to apply the modifications to the copy. In the process of re-optimisation, existing fluences were deleted and new fluences were optimised, following the new request for an optimal dose distribution. When completed, these new fluences were saved to the database. As the next step in the generation of the new plan, final calculations were performed. In this step, the MLC motion control points were generated in order to guide the MLC motion and hence achieve the desired dose distribution during IMRT. This was performed correctly, and a new DRR was also obtained.

(123) The new treatment plan was now complete, and the final step was to save the plan to the database. A 'save all' process started. The items to be saved were: (1) the newly generated actual photon fluence data, (2) the new DRR, and (3) the new MLC control points. When saving the items to the database, the data are first sent sequentially to a holding area on the server. Until all items have been received in this holding area, they will not be saved permanently in the database.

(124) The actual fluence data were saved to the holding area, but a problem occurred when the DRR was being saved. An error message appeared on the TPS, indicating that the data could not be saved. This 'transaction error message' read: 'Please note the following messages and inform your Systems Administrator: Failed to access volume cache file ⟨C:\Program Files\...\504MImageDRR⟩. Possible reasons are... disk full. Do you want to save your changes before application aborts? Yes. No'. The operator pressed 'Yes', which began a second, separate, save transaction. However, the DRR was still locked into the faulty first save transaction; as such, the second save transaction was unable to complete the process, and the software appeared to be 'frozen'.

(125) In reconstruction of the event, it appears that the operator then tried to terminate the software application manually, either by pressing ctrl-alt-del or through Windows Task Manager. This manual termination would have caused the database to perform a roll-back to the last valid state, which in this case contained the newly created actual fluence data, as saved in the first save transaction, and an incomplete part of the newly created DRR from the first save transaction. However, since the

saving of items is sequential, there was no file containing the MLC control points, which should have been saved after the DRR.

(126) A few seconds later, the operator called up the patient's treatment plan on another TPS workstation. Since the new fluence had been saved, the operator was able to calculate the new dose distribution and save this. This could be done regardless of the fact that no MLC control point data had been saved.

(127) According to the quality control procedure in the clinic, the next step should have been to produce a new verification plan and perform in-phantom measurements to verify the consistency of the dose distribution achieved at irradiation with that planned. Also according to the quality control procedure of the facility, a physicist should have reviewed the new treatment plan prior to irradiation of the patient. The verification plan was not calculated at this time, and it is unclear if a physicist reviewed the plan independently. Had these steps been taken, it would have been noticeable that the irradiated area outline lacked an MLC shape, both in the TPS and on the treatment console, resulting in treatment with an open field.

(128) The patient was treated with the incorrect plan (i.e. an open field) for three fractions. Due to the higher number of MUs that the MLC-shaped field would have required, the overdose due to treatment with an open field was substantial. The patient received 39 Gy in three fractions to the head and neck area.

(129) A verification plan was created after the third fraction. When this plan was tested on the treatment unit, the absence of the MLC became apparent and the accident was revealed.

Discussion and lessons. (130) There are four stages at which a problem of this type can be detected: (1) when planning a new treatment configuration, the plan can be inspected to check the dynamic MLC option; (2) an independent review of the plan by a second dosimetrist or a physicist could be expected to detect this type of error; (3) when preparing for the first day set-up, the treatment console would show that the unit was not doing what was expected (i.e. no MLC-leaf movement for an IMRT patient); and (4) a verification plan would definitely discover the wrong dose and dose distribution. A computer 'crash' is not an uncommon event, but can be very dangerous in radiation therapy treatment planning and delivery. Analysis of the potential effect of a computer crash needs to be integrated in the safety assessment, and a procedure needs to be in place for radiation therapy staff to verify the integrity of data systematically after a computer crash.

(131) In summary, even when a quality management system exists, it can be rendered ineffective if alertness and due thought are relaxed, and quality control procedures are ignored in some situations, particularly unusual ones. It is also possible that staff apply the procedures correctly for new treatment plans, but fail to do so for a change in a treatment plan. To minimise the occurrence and impact of this type of problem, radiation oncologists responsible for radiation therapy departments and hospital administrators need to provide continuous encouragement to 'work with awareness' and supervise compliance with procedures, not only for the initial treatment plan but also for treatment modifications.

4.2.3. Errors in imaging for radiation therapy treatment planning

Case 6. Reversal of MRI images (NRC, 2007)

(132) When preparing the treatment for a brain tumour at a clinic, an MRI study of the brain was undertaken. Standard practice was to position and scan the patient 'head first' (i.e. entering the scanner with the head first), and then import the scans into the Gammaknife TPS for optimisation of absorbed dose, dose distribution, and treatment geometry. However, for this patient, due to the 'feet first' scan technique being selected in the software of the imaging unit, the right and left sides of the brain were transposed in the MRI images. When importing the images into the TPS, this was not noticed. The subsequent treatment planning thus targeted an incorrect location in the brain. As a result, the patient received a high radiation dose to the wrong side of the brain.

Discussion and lessons. (133) It appears that the imaging staff performing the MRI scan were not aware of the requirement for an accurate scanning and recording protocol, including image orientation, when imaging for the radiation therapy department. There are two measures to minimise the probability of this type of error: (1) to have clear instructions visibly posted in the MRI suite with written protocols known to and followed by the imaging staff when imaging for radiation therapy treatment planning; and (2) to include procedures in the quality control programme for verifying 'left from right' of safety critical images (e.g. by using fiducial markers where appropriate).

4.2.4. Treatment set-up errors from virtual simulation markers

Case 7. Confusing set-up markers and tattoos when introducing virtual simulation (RO-SIS, 2008)

(134) At the early stages of introducing virtual simulation into a clinic, a breast cancer patient was undergoing this new simulation technique. The intention was to simulate a standard two-field tangential isocentric treatment set-up. The personnel were used to conventional simulation, where this isocentre is determined at the time of simulation. In virtual simulation, the treatment isocentre location is not known at the time of scanning; therefore, set-up tattoos, marked on the patient's skin during virtual simulation, were meant to indicate the origin of the CT co-ordinate system (reference point). In the subsequent treatment planning, the offset from the CT origin reference point to the isocentre of the treatment was determined and noted in the patient's treatment chart.

(135) When the patient came for the first treatment session, the staff at the treatment unit misunderstood what the tattoos indicated, and thought that they marked the treatment isocentre (instead of the CT reference). As a result, the patient was treated in a couch position 3 cm below that intended.

(136) The treatment procedure at this hospital indicated that a check of the source-to-skin distance should be performed. The different distance to the breast with the

patient shifted 3 cm in the longitudinal direction should have been noticed, but this verification was not done. Electronic portal images of the field placements were taken at this first fraction. These were compared with DRRs and approved by a physician; however, the physician was not used to seeing DRRs as reference images. The treatment procedure required the physicist to compare the couch axial position from the treatment plan with the actual couch position in the treatment room at patient set-up, but this was overlooked.

Discussion and lessons. (137) The radiation therapy technologists were not familiar with the different meaning of the tattoos used for virtual simulation, and misunderstood them. A lesson from this event is that new procedures, such as those for virtual simulation, need sufficient training, including exercises for all relevant staff groups, thus providing reasonable assurance that the procedures are understood and critical aspects are fully appreciated. As with any new technique, when introducing virtual simulation, it is important to follow the quality control procedure rigorously. In this case, omitting the source-to-skin distance check at treatment and verification of the axial couch position made this accident more likely to happen.

4.2.5. Digitally reconstructed radiograph errors

Case 8. Geometrical distortion of digitally reconstructed radiographs (Nucletron, 2007)

(138) DRRs are often created in the treatment planning process for use as reference images for the intended set-up of the patient in relation to the treatment unit co-ordinate system and beam placement. The geometric integrity of reference images is of great importance. At the same time, the underlying algorithms for DRR reconstructions in a TPS are difficult for the clinical end-user of the system to verify.

(139) A specific TPS used several methods in parallel for the creation of DRRs. For one of these methods, which was introduced with an updated version of the treatment planning software, an error resulted in incorrect formation and display of the DRR when certain conditions were fulfilled. The problem originated in an error in how the information from the CT slices was loaded into the graphics memory of the TPS computer. This error caused the volume to be stretched out in the z-axis in comparison with the scale of the actual CT series. As a result, there could be a positioning error equal to the minimum distance between slices in the CT series, or in some cases, up to twice this distance.

(140) Since the DRRs exported by the TPS system exhibited the same problem if they were created using the faulty method, the error would then propagate through to treatment delivery. In particular, if an incorrect DRR was used as the reference image, the geometric set-up of the patient would also be incorrect.

Discussion and lessons. (141) A software update (in this case, involving DRR images) is as important as new software or new equipment, and should be tested thoroughly in the factory and properly commissioned at the hospital. Manufacturers can lower the probability of delivering faulty software by performing stringent software tests which challenge the system in a systematic way. Hospitals need to select, plan, and perform a subset of the relevant commissioning tests on the TPS and data transfer. However, a problem that only appears occasionally, when certain conditions are fulfilled, tends to escape tests and verifications. This type of problem should be shared among users and manufacturers, and lessons and commissioning approaches should be disseminated in a timely manner. Dissemination methods include information bulletins from manufacturers, moderated networks, and expert panels.

4.3. Events related to patient data management

4.3.1. Errors when using R&V systems

Case 9. Incorrect manual transfer of treatment parameters (SMIR, 2006; Mayles, 2007; Williams, 2007)

(142) In May 2005, the R&V system at a hospital was upgraded to a more comprehensive electronic patient data management system. Previously, the transfer of treatment parameters had been performed manually. After the upgrade, the system could perform these transfers electronically. This was implemented for most, but not all, treatment procedures in the clinic.

(143) Towards the end of 2005, a young patient came to the hospital with a relatively rare brain tumour (pineoblastoma), and it was decided to give this patient a radiation treatment to the whole central nervous system (CNS). The absorbed dose prescribed was 35 Gy in 20 fractions to the whole CNS, followed by 19.8 Gy in 11 fractions to the site of the tumour (the brain). The craniospinal treatment consisted of two lateral fields covering the brain, matched with an upper and a lower spine field. This type of treatment was considered to be complex, and it was only performed approximately six times per year at this clinic.

(144) As part of the quality management in the clinic, dosimetrists[6] were categorised into five categories ranging from the most junior to the most senior. At the same time, treatment plans were categorised into five categories ranging from the simplest to the most advanced. Contrary to approved procedures, a junior dosimetrist was given the task of developing this advanced treatment plan, with the opportunity to be supervised by a more senior dosimetrist when creating the plan. There are indications that this supervision was reactive rather than active, mainly consisting of responding to queries from the person being supervised. The junior dosimetrist did not have queries for the supervisor and seems to have been unaware of some of the complexities in the plan.

---

[6] The meaning of the word 'dosimetrist' is not uniform throughout the world; it is used in this report to mean the person who performs the treatment planning and clinical dosimetry.

(145) With the old procedure, the planning system calculated the number of MUs needed to give an absorbed dose of 1 Gy at the normalisation point. These MUs were subsequently scaled up to the prescribed dose by manual multiplication with the dose per fraction. With the new procedure for automatic electronic data transfer to the data management system, the TPS provides the number of MUs required to give the prescribed dose to the dose prescription point. The craniospinal treatment technique was one of the few techniques that had not yet been included in the new procedure of automatic electronic transfer. Instead, the treatment planning should have been performed according to the old procedure.

(146) When planning the left and right lateral fields covering the brain in the treatment of this particular patient, the junior dosimetrist let the TPS calculate the MUs for the prescribed dose (i.e. the new procedure) instead of the MUs to deliver 1 Gy (i.e. the old procedure). The junior dosimetrist then transferred this MU setting to the manual planning form, which was passed to a radiation therapy technologist for manual upscaling calculations of the MUs. This manual planning form contained the MUs corresponding to the prescribed dose, instead of 1 Gy. The technologist performed the scaling up of the MUs according to the old procedure, which meant that the number of MUs would be 75% too high for each of the lateral head fields. It should also be noted, however, that the dosimetrist made a second error when filling out the MU settings in the manual planning form, in that the figure for the total number of fractions was wrong in the calculations, leading to a 67% overdose, instead of 75%, according to the ratio of 1.75 Gy to 1 Gy. These errors were not found by the more senior planner who checked the calculations.

(147) In the resulting treatment, the patient received 2.92 Gy per fraction to the head, instead of the intended 1.75 Gy. This continued for 19 fractions, when the same junior dosimetrist committed the same error with another plan. This second error was spotted when the new treatment plan was checked, and the original error was found. The patient died 9 months after the accident.

Discussion and lessons. (148) The initial reflection to make is that using two different methods to transfer data, with the associated opportunity for errors, should be avoided as much as possible. If there are strong reasons to keep the manual transfer of treatment parameters for some treatment procedures, and if a type of treatment is only performed a few times per year, it may be sensible to always assign the treatment to the same person (or two people). There is a challenge for hospital administrators to ensure a working environment that facilitates alertness, due thought, and compliance with procedures. Relaxation of compliance with procedures seems to have occurred at two different levels at this hospital: (1) assignment of an advanced task to a junior planner, contrary to hospital procedures; and (2) failure of a calculation check by the senior dosimetrist. An important lesson is that this occurred in spite of the quality management system in place, with a sophisticated five-level staff structure and five-level categories for treatment plans.

4.4. Events related to treatment delivery and treatment verification

4.4.1. Significant radiation exposure from electronic portal imaging

Case 10. Excessive exposure by the daily use of electronic portal imaging (Derreumaux et al., 2008)

(149) Since the introduction of EPIDs, clinics have been able to monitor the set-up of individual patients more readily than with portal films. While a modern EPID can display a portal image with low exposure of the patient (typically a few MUs), some of the earlier EPID models required much higher exposure to form an image.

(150) A clinic (the same one as in Case 4) installed an EPID that was based on a matrix of liquid ionisation chambers, and it was decided that the set-up position of patients treated for prostate cancer should be verified daily. In order to do this, they took two single-exposure images for each patient (i.e. one anterioposterior and one lateral image) on each treatment day. In addition to these daily patient positioning images, weekly portal images were taken for each patient and for all fields to confirm the correct placement of the irradiation fields. The weekly portal images were performed using a double-exposure technique (one exposure for the shaped field and one for the open field) where the same anatomical parts were exposed twice.

(151) As part of the clinic's practice, when performing the weekly double-exposure verification of the treatment fields, the MUs used for the shaped field, but not the open field, were deducted from the MUs to be used for treatment, with the intention of giving the prescribed absorbed dose on that treatment day. However, neither the increased exposure arising from the daily patient positioning images, nor the open field element of the weekly double-exposure verification of the treatment fields were considered in the recording of the total dose given to the patient.

(152) The EPID in the clinic required a relatively high exposure to obtain an electronic portal image. As a result, it has been estimated that each patient received a daily absorbed dose of between 0.15 and 0.20 Gy in excess of the prescribed dose due to the electronic portal imaging protocol. In total, 397 patients were affected between 2001 and 2006, and received an absorbed dose of 8–10% higher than intended. All patients affected by the overdose in Case 4 were also affected by this error, thus adding to the already substantial overdose of these patients.

Discussion and lessons. (153) There seems to have been a lack of awareness of the magnitude of the additional absorbed dose resulting from frequent use of the imaging system used for set-up, and the fact that this dose could be significant with respect to the total prescribed treatment dose. This lack of awareness seems to be the reason why the radiation dose from daily imaging was not taken into account. Before introducing new imaging technologies and verification procedures into clinical practice, the additional radiation exposure needs to be assessed.

4.4.2. Errors with stereotactic radiosurgery field size

Case 11. Incorrect field size used for stereotactic treatment (Derreumaux et al., 2008)

(154) A clinic was using a linear accelerator for stereotactic treatment of intracranial targets using a set of additional cylindrical collimators with opening diameters ranging from 10 to 30 mm, mounted on an opaque brass supporting tray that was attached to the accelerator's accessory holder. For the correct use of these cylindrical collimators, it was necessary to set the jaws to give a rectangular collimator aperture of 4 cm × 4 cm.

(155) When treating a patient with arteriovenous malformation with a single fraction, the additional cylindrical collimator for stereotactic treatment was attached to the linear accelerator. The operator was verbally instructed by the physicist to narrow the collimator aperture to '40, 40', but instead of using the field size '40 mm × 40 mm' as intended, the operator used the field size '40 cm × 40 cm'.

(156) As a consequence of this communication error, the fully opened field was applied to the patient through the brass tray supporting the additional cylindrical collimator. As the brass supporting tray would only result in very limited attenuation of the beam, nearly the full absorbed dose was given to large areas outside the target volume.

(157) When evaluating the magnitude of overexposure locally, the impact was underestimated. This led to incorrect assessment of the severity of the clinical consequences, which were thus not fully appreciated or provided for.

(158) The clinical consequences attributed to this were fibrosis and oesotracheal fistula, which led to a surgical intervention and subsequently death of the patient from haemorrhage. The very severe clinical consequences do not seem to match the dose calculated for this volume, suggesting the potential presence of an additional problem.

Discussion and lessons. (159) The following reflection can be made from this event: when the operator heard '40, 40', he/she associated it with a conventional radiation therapy field of 40 cm × 40 cm, instead of thinking of radiosurgery. Furthermore, a cylindrical collimator of a few millimetres diameter inside the 40 cm × 40 cm beam did not appear strange to him/her and it did not trigger a question. Thus, the operator seems to have been performing a new technique – stereotactic radiation therapy – without full understanding. It is not clear whether or not there were well-documented prescription and treatment procedures, given the fact that the instruction on the field size was verbal, and its misinterpretation caused the accidental exposure. In summary, a combination of insufficient training in the new technique, and verbal instruction instead of a written communication of essential features of the treatment plan caused this accidental exposure.

4.5. References

Ash, D., 2007. Lessons from Epinal. Clin. Oncol. 19, 614–615.

ASN, 2007. Report Concerning the Radiation Therapy Incident at the University Hospital Centre (CHU) in Toulouse – Rangueil Hospital. Autorité de Sûreté Nucléaire, Bordeaux.

Derreumaux, S., Etard, C., Huet, C., et al., 2008. Lessons from recent accidents in radiation therapy in France. Radiat. Prot. Dosimetry 131, 130–135.

IGAS, 2007. Summary of ASN Report No. 2006 ENSTR 019 - IGAS No. RM 2007-015P on the Epinal Radiation Therapy Accident. Wack, G., Lalande, F., Seligman, M.D. (Eds.), Autorité de Sûreté Nucléaire and Inspection Générale des Affaires Sociales, Paris.

Mayles, W.P.M., 2007. The Glasgow incident – a physicist's reflections. Clin. Oncol. 19, 4–7.

NRC, 2007. Gamma Knife Treatment to Wrong Side of Brain. Event Notification Report 43746. US Nuclear Regulatory Commission, Washington, DC.

Nucletron, 2007. Incorrect Creation of DRR Using GPU-based Methods. Customer Information Bulletin (Nucletron) CIB-OTP 192.047-00. Nucletron, Veenendaal.

NYC Department of Health and Mental Hygiene, 2005. ORH Information Notice 2005-01. Office of Radiological Health, NYC Department of Health and Mental Hygiene, New York, NY.

Rosenwald, J.C., 2007. Personal communication.

ROSIS, 2008. Internet-based Radiation Oncology Safety Information System. Available at: http://www.rosis.info.

Saint-Luc Hospital, Brussles, Belgium, 2008. Internal Report. Saint-Luc Hospital, Brussels.

SMIR, 2006. Accidental overexposure of patient Lisa Norris during radiation therapy treatment at the Beatson Oncology Centre, Glasgow in January 2006. Report of an Investigation by the Inspector Appointed by the Scottish Ministers for the Ionising Radiation (Medical Exposures) Regulations 2000. Scottish Executive, Edinburgh.

VMS, 2005. [Treatment Facility] Incident Evaluation Summary. CP-2005-049 Varian Medical Systems, Palo Alto, CA, pp. 1–12.

Williams, M.V., 2007. Radiation therapy near misses, incidents and errors: radiation therapy incident in Glasgow. Clin. Oncol. 19, 1–3.

5. PROSPECTIVE APPROACHES TO AVOIDING ACCIDENTAL EXPOSURES

(160) While valuable lessons can be learnt from the detailed analysis of incidents and accidental exposures which have occurred, these lessons are necessarily limited to reported experience. Undoubtedly, there are potential incidents which have not yet happened but which are possible, and actual incidents which have not been openly reported. Re-occurrence of these types of incidents can only be avoided if they are anticipated. Furthermore, increased complexity in the clinic places new demands on the selection of quality control checks. The approach of following all-inclusive lists of tests and all possible controls may become impractical within the context of limited time and resources. Therefore, particularly with technological and process changes in radiation therapy, retrospective approaches are not sufficient and all-inclusive quality control checks may not be feasible. There is, thus, a need for prospective, structured, and systematic approaches to the identification of system weaknesses and the anticipation of failure modes, including the evaluation and comparison of potential risks from each identified failure mode. Such approaches should allow rational selection of the checks to be performed, and facilitate the distribution of resources in a manner likely to be most beneficial to the patient.

5.1. Treatment process trees

(161) The identification of weaknesses in the system requires an understanding of the system itself. A helpful approach to the understanding of the system is through visualisation by means of a process flow diagram or map. The generic process through which a patient passes in any encounter with a healthcare system includes the following five steps: diagnostic assessment, treatment prescription, treatment preparation, treatment delivery, and follow-up (Ekaette et al., 2006). Patient data flow from one step to the next with return loops as necessary. A feature which characterises modern radiation therapy is an electronic RTIS which may form part of the electronic medical record, and which links all or most of the processes involved in the five steps above. It is the availability of such systems capable of transferring large amounts of data that has enabled the introduction of new treatment strategies such as IMRT and image-guided radiation therapy in the clinic.

(162) The intermediate steps of prescription, preparation, and delivery of treatment are of most interest in this report. Each of the major steps can be divided into substeps. For example, preparation for treatment includes patient immobilisation, image segmentation and structure delineation, calculation of 3D dose distributions and machine settings (e.g. MUs), and the transfer of data to the RTIS. Should it be useful for the analysis, these substeps may be further subdivided. For example, the dose calculation substep includes the specification of objectives and constraints used in optimisation of the plan. Examples of a process flow chart for the three intermediate steps in radiation therapy have been presented by Rath (2008) and Ford et al. (2009).

(163) An alternative representation of the activities taking place prior to and during a course of radiation therapy is known as a 'process tree'. The trunk of the tree conducts the patient from entry into the system to successful completion of the treatment. The boughs connected to the trunk represent the tasks, such as immobilisation, image segmentation and structure delineation, and treatment planning, which are necessary for completion of the treatment. Along each bough, the substeps, such as choice of fusion and segmentation algorithms and margin selection, are identified. An example of a process tree for high-dose-rate brachytherapy has been presented by Thomadsen et al. (2003).

(164) The clinical processes, visualised by means of the clinical process flow diagram or clinical process tree, are carried out using the clinical infrastructure of the institution (hardware, software, documentation, etc.). Establishing and maintaining the clinical infrastructure of an institution also involves acceptance testing, calibration, commissioning, and regular quality control tests. Process maps and trees describing activities related to the maintenance of the clinical infrastructure are also useful in the identification of failure modes. Past experience indicates that failure modes related to the infrastructure generally have the most significant consequences, as their effects are systematic and can affect large numbers of patients. Examples include errors in beam calibration with catastrophic consequences (IAEA, 1998, 2000, 2001; ICRP, 2000).

(165) The importance of acknowledging the possibility of failures of both infrastructure and processes relating to the use of new technologies can be seen from the case histories in Section 4. Case 1 shows that although each step in the patient treatment process may be carried out correctly, it is possible that the infrastructure is faulty due to an error in calibration. Similarly, it is possible that even when each relevant component of the clinical infrastructure is performing as expected, the clinical process can be flawed, with severe consequences for a large number of patients (Case 4).

(166) Clinical process maps and trees are graphical representations, broadly in chronological order, of the activities to be performed for the successful completion of treatment of a patient. They may also be inferred from classification schemes used for clinical incident reporting and analysis; for example, 'Towards Safer Radiation Therapy' (Royal College of Radiologists et al., 2008). As mentioned above, it may be helpful to separately construct an infrastructure process tree which describes the calibration, commissioning, and regular quality control of equipment, maintenance, and release of clinical data and procedures, together with any other components required for the treatment of all or a cohort of patients.

(167) The process tree developed for use in an individual clinical programme should reflect the structure of that programme and the sequence of activities taking place in a manner that is logical and clear to the multidisciplinary team responsible for the care of the patient. A balance needs to be struck between simplicity, to maintain comprehensibility, and complexity, to capture all possible failure modes of the system. An approach to the validation of local process maps and trees is the assessment of their ability to capture all the historical incidents described in this and other relevant documents as far as they are applicable to local circumstances. This is a

necessary but not sufficient condition for the adequacy of representation. In the case of new technologies and treatment strategies, additional processes over and above those identified from historical events will need to be incorporated.

5.2. Process tree and the design of a quality management system

(168) Process flow illustrations can facilitate the design of quality management programmes. Each process or group of processes should be recognised within the quality management programme. Key components of a quality management system in radiation therapy are the commissioning and recommissioning of infrastructure and clinical processes, as well as regular quality control. The quality management system needs to encompass infrastructure (e.g. monthly checks of a linear accelerator), clinical processes (e.g. checks of MUs calculated for individual treatments), and, increasingly, patient-specific activities (e.g. experimental verification of the fluence distribution for IMRT beams). An approved process flow illustration is helpful in the design of a comprehensive, effective, and efficient quality management system.

5.3. Failure modes

(169) Process trees facilitate the next step in a prospective approach to risk management, which is the identification of possible failure modes at each step in the process. As this is a prospective approach, prior statistically valid experience is likely to be limited. The approach normally adopted for the identification of failure modes is to convene an expert panel which reviews the process flow illustrations and, within a structured context, uses their judgement to prepare a list of possible failure modes. There are two major challenges in completing this step. The first is to be confident that all possible significant failure modes have been identified. In meeting this challenge, the range of experience of the expert panel will be important. As a minimum, possible failure modes should include those reported in the literature and on public databases. The second challenge is describing the failure modes unambiguously. Not only must the description of the failure mode be completely clear, but also if the same failure mode can occur in two different steps of the treatment process, it should be differentiated as two separate failure modes. Different possible sources or root causes need to be differentiated. These could arise either from equipment malfunction or human error, and need to be described as such. The failure mode and even the clinical outcome might be the same, but the likelihood of occurrence and, importantly, remedial measures could be quite different. The fault tree description of failure modes can be useful in this regard (Thomadsen et al., 2003).

(170) To be useful, the description of the failure modes should be sufficient to guide any changes in the treatment process or in the quality management programme that result from the analysis. Having identified potential failure modes on the basis of global experience, and ideally with input from equipment manufacturers, the next step is to assess the risk associated with each failure mode.

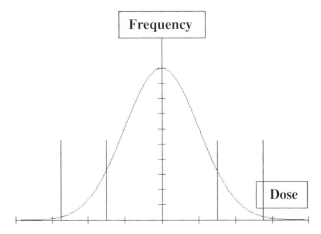

Fig. 5.1. Frequency distribution of the ratio of delivered to prescribed dose in the absence of systematic errors.

5.4. Risk

(171) For the purposes of this report, risk can be regarded as a function of the probability of an event occurring and the severity of the consequences for the patient should the event occur. Severity indices may also reflect the numbers of patients involved.

(172) Fig. 5.1 illustrates how risk might be assessed. It represents an idealised frequency distribution of delivered doses when no systematic errors are present, with the central peak indicating the radiation oncologist's prescribed dose. Some variability in the relationship between the prescribed and delivered doses is inevitable due to, for example, uncertainties in the absorbed dose determination from the output of the treatment machine, algorithm limitations in the TPS, and patient positioning reproducibility over a course of treatment requiring 30 or more fractions. Using clinical data and observation, it is possible to determine a range of acceptability for the delivered dose (Mijnheer et al., 1987). At some deviation from the prescription, the treatment becomes unacceptable in terms of the negative consequences for the patient. Although overdosing often attracts more attention, inadvertent underdosing can also have major consequences for the patient. The thresholds at which catastrophic clinical events occur will depend on the clinical situation. However, in order to track significant accidental exposures, some regulators have defined the threshold to be at a specific value over a course of treatment (e.g. the US Nuclear Regulatory Commission). The region between acceptable and unacceptable treatments could be described as 'suboptimal'. There may not be clear clinical evidence that treatments falling in that region result in deleterious effects for the patient. However, significant departures from (evidence-based) prescription are clearly not desirable.

(173) Fig. 5.1 describes the situation in which there are no systematic errors causing dose deviations from prescription, such as an error in calibration (Case 1) or use

of the technology (Case 4). When systematic deviations are present, the curve will no longer be centred on the oncologists' prescription, and hence the frequency of either under- or overdosing will be increased. If there is a systematic error affecting all patients, the whole curve would shift to the right (overdose) or left (underdose) by the amount of the deviation (see Fig. 5.2). There may also be a systematic error that only affects certain types of treatments and therefore only affects a limited cohort of patients (Fig. 5.3). The end result of a systematic deviation is that a larger number of patients exceed the threshold of acceptability, possibly with serious consequences (ICRP, 2000).

(174) In the figures, the abscissa could be 4D dose, i.e. including fractionation in time. A distribution could also be constructed to reflect inadequacy in following the oncologist's prescription for the volumes to be irradiated and, equally important, the volumes not to be irradiated. It is noted that volume in this context is also 4D as its location within the patient may change with time due to, for example, respiration. One of the newer technologies under active investigation is gated therapy for lung cancer. Loss of synchrony between motion of the target and time of irradiation can have severe deleterious effects. Enhancing safety is interpreted as reducing the mass density of the distribution in the regions of unacceptability. If the distribution is not centred on the oncologist's prescription, systematic effects are present and must be rectified. If the distribution is Gaussian or close to Gaussian, shrinking the tails might also be expected to narrow the peak, corresponding to reduced variability and hence an improvement in quality for a centred distribution.

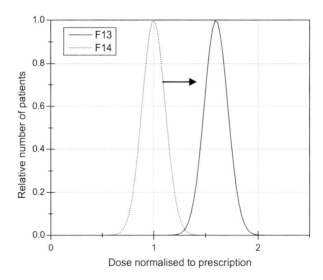

Fig. 5.2. Frequency distribution of the ratio of delivered to prescribed dose in the presence of a systematic deviation causing overexposure of all patients. This example is similar to the event that occurred in Costa Rica due to a beam calibration error (IAEA, 1998). The dotted line shows normal treatment in the absence of an accidental exposure and the solid line shows the actual doses from the accidental exposure.

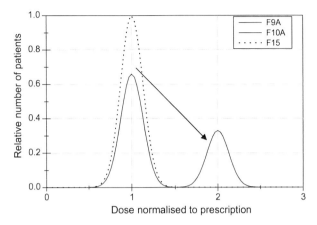

Fig. 5.3. Frequency distribution of the ratio of delivered to prescribed dose in the presence of a systematic deviation affecting approximately one-third of patients. This example is similar to the event that occurred in Panama due to modified use of the treatment planning system in which only some patients treated in the abdominal region were affected (IAEA, 2001). The dotted line shows normal treatment in the absence of an accidental exposure, while the solid line on the right shows the doses to patients affected by an accidental over exposure and the solid line on the left corresponds to the doses to the rest of the patients not affected by the accidental exposure.

5.5. Three prospective approaches

(175) Once the failure modes have been identified, the task becomes that of assessing the probability of an unacceptable event occurring (the first contribution to the ordinate in the diagram), assessing the severity or consequences of the event should it occur (the abscissa), and in some of the approaches, assessing the likelihood that the event, should it occur, will not be detected during quality control checks (the second contribution to the ordinate) and will hence have a negative impact on the patient's treatment. The prospective approaches of risk assessment described below employ methods of analysing the problem in this way.

(176) The three prospective approaches which are most commonly used are failure modes and effects analysis (FMEA) (Stamatis, 1995), probabilistic safety assessment (FORO, 2009; IAEA, 2006; Ortiz López et al., 2008a; Vilaragut Llanes et al., 2008), and risk matrix (Duménigo et al., 2008; Ortiz López et al., 2008b). They are not totally independent, as FMEA is often used as the first step to probabilistic safety assessments, as described later in this report.

5.5.1. Failure mode and effect analysis

(177) Examples of the application of an FMEA to radiation therapy are those performed by Task Group 100 of the American Association of Physicists in Medicine (Huq et al., 2008) and Ford et al. (2009). Three numerical values were used to describe each failure mode. O (for occurrence) describes the probability that a particular untoward event will occur. S (for severity) is a measure of the severity of the

consequences resulting from the failure mode if it is not detected and corrected. D (for detectability) describes the probability that the failure will not be detected before the treatment commences or the failure is effective. In Task Group 100's implementation, O ranges from 1 (failure unlikely, <1 in 10^4) to 10 (highly likely, >5% of the time), S ranges from 1 (no danger, minimal disturbance of clinical routine) to 10 (catastrophic if persists through treatment), and D ranges from 1 (very detectable, $\leqslant$0.01% of events go undetected throughout treatment) to 10 (very hard to detect, >20% of failures persist through the treatment course). An important point to note in the evaluation of D is that the failure mode is assumed not to have been detected through regular quality control checks within the subprocess where the failure occurred. Thus, in Task Group 100's implementation of FMEA, the likelihood of (lack of) detection at any point further downstream from the subprocess in which the failure occurred is estimated. Ford et al.'s implementation is similar but differs in some of the details of the grading of O, S, and D (Ford et al., 2009).

(178) Multiplying these three numbers together yields a risk priority number (RPN) which can be used for prioritising quality control tests and activities.

(179) The presentation of a complex and comprehensive FMEA is challenging. An instructive approach is to incorporate the findings of the FMEA within the process flow diagram or process tree which was developed at the initiation of the analysis. For example, failure modes with RPNs greater than a certain threshold might be highlighted in a process tree. This diagram then contains the essential elements of a fault tree. Fault trees can be constructed using either or both retrospective and prospective analyses, and are therefore a flexible and useful tool in safety analysis.

(180) The calculation of RPNs implies knowledge of the functional relationship between probabilities of occurrence and detectability, and the severity of the consequences in quantifying risk. Institutional judgement may dictate that more serious events, even with a relatively low probability of occurrence, deserve increased attention. Such a policy can also be incorporated into the fault tree by displaying failure modes with a severity that exceeds a certain threshold, irrespective of the probability of occurrence/detectability.

5.5.2. Probabilistic safety assessment

(181) Probabilistic safety assessment is a prospective tool that has been successfully used in the aeronautics, nuclear, and petrochemical industries, and which has also been proposed for use with radiation sources in industry and medicine (ICRP, 1997). More recently, probabilistic safety assessment has been applied to radiotherapy treatment with linear accelerators (Ortiz López et al., 2008a; Vilaragut Llanes et al., 2008) by a task group of the Ibero American FORO of Nuclear and Radiation Safety Regulatory Agencies (FORO, 2009). The task group was multidisciplinary, involving radiation oncologists, medical physicists, technologists, regulators, maintenance engineers, and specialists in probabilistic safety assessments. The study was devoted to conventional treatments with accelerators, excluding the infrastructure elements such as commissioning of the facility, which was left for a separate study.

(182) FMEA of equipment was used to obtain a list of possible failure modes. Since every failure mode is an initiator of an event sequence, it may evolve towards an accidental exposure. These event sequences are graphically modelled by 'event trees'. The accidental exposure may or may not occur, depending on the success or failure of existing safety provisions meant to stop the event sequence. The risk of accidental exposure is obtained as a combination of the numerical value of the frequency of the initiating event and the probability of failure of the safety provisions, using the event trees.

(183) To perform this calculation, data on numerical frequencies of the initiating events and the probability of failure of each of the safety provisions are needed. When frequency data of the relevant failure modes are available, the quantification can proceed, in principle, but when frequency data for a failure mode are not available, it is necessary to breakdown the specific failure mode into more basic failure modes, the frequency of which are known from available generic databases. An example of a basic failure mode is the rate of errors made by a person when copying numbers from one place to another.

(184) Sometimes, even when data for a failure mode are available, investigation of the impact of certain specific safety provisions or smaller components on the total is needed, in which case the breakdown can continue in order to capture this information. The different degree of specificity does not mean different numerical results in terms of risk, but simply enables more detailed information to be obtained. Failure modes that are too general may lead to a loss of information, and failure modes that are too detailed may become impractical. The degree of specificity to be chosen is therefore an optimisation exercise.

(185) Breakdown of the more general failure mode into basic failure modes is modelled by a so-called 'fault tree'. The tree represents the path between the general and basic failure modes. This very systematic methodology imposes discipline on analysts and constrains their subjectivity, while facilitating the conduct of the study.

(186) In the FORO's study referred to above, accident sequences were graphically modelled by means of fault trees and event trees. After formulating the Boolean relations of these trees, a Boolean reduction was conducted by computer software to obtain the minimum combination of equipment faults and human errors (referred to as 'minimal cut sets') which produce a given accidental sequence. Quantification is then performed by calculating probabilities using the trees of minimal cut sets.

(187) Given the scarcity of statistical data on reliability of equipment and human error in radiation therapy that can be used in the quantification process, generic databases from several sources (IAEA, 1988, 1997; US Department of Energy, 1996) were used to estimate the reliability of equipment. This approach is typically recommended for topical probabilistic safety assessments that are applied for the first time.

(188) In a similar way, screening values were used for the probability of human error, i.e. conservative values which allow the most important human actions to be filtered, and focusing efforts on these actions in further detailed analysis. Although screening values increase uncertainty, the approach is adequate for relative comparisons since the whole quantification was performed using the same type of self-consistent data. Comparative calculations are useful for evaluating the relative

contribution of various event sequences to the global risk, analysis of importance, sensitivity analysis, and evaluation of risk reduction from different safety measures.

(189) Detailed discussion of the findings is beyond the scope of this report, but major general findings can be summarised. With an FMEA performed as part of a probabilistic safety assessment, 443 failure modes were identified. Without rational screening of this data, managerial decisions to prioritise and allocate staff time and resources from the long list of failure modes would be impractical. For this purpose, the Boolean evaluation of fault trees provides a tool for an analysis of importance of the event sequences, resulting in a rational prioritisation. Three major findings are given below:

- Quantification of the risk shows that a few event sequences are responsible for most of the accidental exposures that originate in the treatment process. As an example, the FORO's study found that a few event sequences are responsible for 90% of the total risk of the potentially catastrophic accidental exposures involving multiple patients. This finding is crucial for risk-informed and cost-effective decisions. Since this particular study was devoted to the treatment process alone, beam calibration and commissioning being excluded, accidental exposures from calibration errors were not covered in this report.
- Accidental exposures of single patients are much more likely to occur than more catastrophic multiple-patient exposures. Attention has primarily focused on events of a catastrophic nature with low probability. The higher risk of accidental exposure of a single patient has not attracted so much attention, probably because single-patient events are more likely to be under-reported. Safety measures to avoid single-patient events deserve more attention.
- The 'analysis of importance' of a given failure mode estimates the risk increase caused by changing the assumed probability of occurrence to 100%, and the 'analysis of importance' of an additional safety measure estimates the risk reduction if that measure were implemented. This facilitates identification of the failure modes with the largest risk increase should they occur with certainty, and additional safety measures which would lead to the largest risk reduction if they were implemented. Examples of types of findings with the greatest impact are: (1) the need for more specific safety requirements on developing and testing software, (2) specific features of R&V systems, (3) allocation of space for patient and site photographs in electronic treatment sheets, (4) improvement of CT forms and treatment plans, (5) a secondary MU calculation independent from the TPS, and (6) verification of the patient set-up by a second technologist.

5.5.3. Risk matrix

(190) The risk matrix method consists of identifying potential events, applying a simple conservative screening to filter lower-risk events, devoting a more detailed and realistic safety assessment to the fewer higher-risk events, and identifying additional safety measures needed to bring the latter to a low-risk level. The core of the method is an efficient screening approach, called the 'risk matrix', to filtering the

Table 5.1 Complete risk matrix containing all combinations of the four levels of frequency of occurrence of the initiating event (f), the four levels of probability of failure of the set of safety measures (P), and the four levels of severity of consequences (C) of the possible accidental exposure, along with the resulting level of risk (R).

f_H	P_H	C_{VS}	R_{VH}	f_H	P_H	C_S	R_{VH}	f_H	P_H	C_M	R_H	f_H	P_H	C_m	R_L
f_M	P_H	C_{VS}	R_{VH}	f_M	P_H	C_S	R_H	f_M	P_H	C_M	R_H	f_M	P_H	C_m	R_L
f_L	P_H	C_{VS}	R_H	f_L	P_H	C_S	R_H	f_L	P_H	C_M	R_L	f_L	P_H	C_m	R_L
f_{VL}	P_H	C_{VS}	R_H	f_{VL}	P_H	C_S	R_H	f_{VL}	P_H	C_M	R_L	f_{VL}	P_H	C_m	R_L
f_H	P_M	C_{VS}	R_{VH}	f_H	P_M	C_S	R_H	f_H	P_M	C_M	R_H	f_H	P_M	C_m	R_L
f_M	P_M	C_{VS}	R_H	f_M	P_M	C_S	R_H	f_M	P_M	C_M	R_L	f_M	P_M	C_m	R_L
f_L	P_M	C_{VS}	R_H	f_L	P_M	C_S	R_H	f_L	P_M	C_M	R_L	f_L	P_M	C_m	R_{VL}
f_{VL}	P_M	C_{VS}	R_H	f_{VL}	P_M	C_S	R_L	f_{VL}	P_M	C_M	R_L	f_{VL}	P_M	C_m	R_{VL}
f_H	P_L	C_{VS}	R_H	f_H	P_L	C_S	R_H	f_H	P_L	C_M	R_L	f_H	P_L	C_m	R_{VL}
f_M	P_L	C_{VS}	R_H	f_M	P_L	C_S	R_H	f_M	P_L	C_M	R_L	f_M	P_L	C_m	R_{VL}
f_L	P_L	C_{VS}	R_L	f_L	P_L	C_S	R_L	f_L	P_L	C_M	R_L	f_L	P_L	C_m	R_{VL}
f_{VL}	P_L	C_{VS}	R_L	f_{VL}	P_L	C_S	R_L	f_{VL}	P_L	C_M	R_L	f_{VL}	P_L	C_m	R_{VL}
f_H	P_{VL}	C_{VS}	R_H	f_H	P_{VL}	C_S	R_L	f_H	P_{VL}	C_M	R_L	f_H	P_{VL}	C_m	R_{VL}
f_M	P_{VL}	C_{VS}	R_L	f_M	P_{VL}	C_S	R_L	f_M	P_{VL}	C_M	R_L	f_M	P_{VL}	C_m	R_{VL}
f_L	P_{VL}	C_{VS}	R_L	f_L	P_{VL}	C_S	R_{VL}	f_L	P_{VL}	C_M	R_{VL}	f_L	P_{VL}	C_m	R_{VL}
f_{VL}	P_{VL}	C_{VS}	R_L	f_{VL}	P_{VL}	C_S	R_{VL}	f_{VL}	P_{VL}	C_M	R_{VL}	f_{VL}	P_{VL}	C_m	R_{VL}

VH, very high; H, high; L, low; VL, very low; VS, very severe; S, severe; M, medium; m, minor.

low-risk events and devoting a focused attention on the higher-risk events. The level of risk is estimated for each initiating event from a logical combination of the frequency of the initiating event, the probability of failure of the safety provisions relevant to this event, and the severity of the consequences. The logical combination is called the 'risk matrix'.

(191) The combination is obtained by a using a four-level scale for the frequency, the probability of failure of the safety provisions, the severity of the consequences, and the resulting risk. The four levels are, for example, very low, low, high, and very high. For the consequences, the four levels of the scale are very severe, severe, medium, and minor.

(192) Once the risk level has been determined by applying the logical combination to each initiating event, further efforts are focused on the event sequences classified by the risk matrix as high or very high risk. This part is used as a screening process to filter out events with negligible risk.

(193) The logic for building the matrix is as follows: two parameters of the same level combine into the same level, i.e. 'low' with 'low' results in 'low'; 'high' with 'high' results in 'high'. Two parameters of different levels combine into an intermediate level: 'high' with 'low' results in 'medium'. When the intermediate level is not defined in the scale for the combination, the most conservative combination is chosen, i.e. 'high' with 'low' results in 'high'. Since three parameters need to be combined to determine the risk, the process has two steps: the level of frequency is combined with the level of probability of failure of the safety provisions, and the resulting combination is further combined with the level of consequences. The complete set of combinations resulting from this logical process is given in Table 5.1. The end result is the level of risk. The outcome of the risk matrix screening consists of a list of potentially higher-risk events. They

are the focus of the subsequent more detailed analysis, which consists of systematically asking the following questions for each of these event sequences: 'How robust are the safety provisions?', 'Can the level of frequency of occurrence of the event or of its consequences be reduced?', 'Is there a need to add one or more safety provisions to reduce the risk to an acceptable level?'. The answers to these questions constitute the conclusions and recommendations from the study.

5.6. Closing the loop and applying prospective methods

(194) The objective of application of any of these prospective analyses is minimisation of the risk of accidental exposure of patients. The methods of prospective analysis discussed above, anchored as they are in process flow diagrams and process trees, suggest how improvements in safety might be implemented. A given failure mode will result in consequences of a given maximum severity. Estimates of the severity of failure modes provide an opportunity for prioritising safety improvement initiatives. The analysis also provides estimates of the likelihood of a failure occurring, and the likelihood of its being detected. If a prospective analysis has identified a failure mode with a high probability of occurrence, this suggests that the activities involved warrant re-examination. Linking the analysis to the clinical or infrastructure process flow illustrations identifies which process or processes are involved. Where a weakness is identified, either prospectively or retrospectively, that process requires revision with the intention of minimising the probability of occurrence. Similarly, failure modes associated with low detectability suggest that the programme of quality control checks needs to be re-examined. With a quality management system built on the foundation of a process flow illustration, the activities which may require closer scrutiny will be more readily identified.

(195) Whether a quantitative (FMEA and probabilistic safety assessment) or a more qualitative (risk matrix) prospective analysis is undertaken, the results will facilitate prioritisation of remedial measures based on the overall estimated risk to the patient, or severity or some other criterion selected by the institution.

(196) It appears from the foregoing that full prospective analyses are complex and time-consuming. National and international organisations and professional bodies can assist individual clinics by recommending easily interpretable scales for quantifying occurrence, severity, and detectability, and by illustrating the use of prospective techniques applied to generic process trees (Huq et al., 2008). Developers of patient safety databases, while primarily enabling retrospective analyses, can facilitate prospective analyses as historical information is useful for validating, in part, the process descriptions upon which prospective analyses are based.

(197) To summarise, prospective analyses are an essential component of a safety assessment, particularly for technological and process changes, and are, in addition, a useful approach to managing risks with existing equipment and current work practices. A properly conducted prospective analysis helps to identify potential failure modes and the severity of the ensuing clinical consequences. Where a particular failure mode is associated with a high risk of occurrence, the infrastructure component or clinical process needs to be revised, redesigned, and implemented accordingly.

When a prospective analysis indicates that a particular failure mode is unlikely to be detected, the quality control checks and associated safety measures need to be strengthened.

5.7. References

Duménigo, C., Ramírez, M.L., Ortiz López, P., et al., 2008. Risk analysis methods: their importance for safety assessment of practices using radiation. XII Congress of the International Radiation Protection Association, 19–24 October 2008, Buenos Aires.

Ekaette, E.U., Lee, R.C., Cooke, D.L., Kelly, K.L., Dunscombe, P., 2006. Risk analysis in radiation treatment: application of a new taxonomic structure. Radiother. Oncol. 74, 282–287.

Ford, E.C., Gaudette, R., Myers, L., et al., 2009. Evaluation of safety in a radiation oncology setting using failure modes and effects analysis. Int. J. Radiat. Oncol. Biol. Phys.74, 852–858.

FORO, 2009. Análisis probabilista de seguridad de tratamientos de radioterapia con acelerador lineal. Ibero American Forum of Radiation and Nuclear Safety Regulatory Organizations. Available at: http://www.foroiberam.org.

Huq, M.S., Fraass, B.A., Dunscombe, P.B., et al., 2008. A method for evaluating quality assurance needs in radiation therapy. Int. J. Radiat. Oncol. Biol. Phys. 71, S170–S173.

IAEA, 1988. Component Reliability Data for Use in Probabilistic Safety Assessment. TECDOC-478. International Atomic Energy Agency, Vienna.

IAEA, 1997. Generic Component Reliability Data for Research Reactor PSA. TECDOC-930. International Atomic Energy Agency, Vienna.

IAEA, 1998. Accidental Overexposure of Radiotherapy Patients in San José, Costa Rica. International Atomic Energy Agency, Vienna.

IAEA, 2000. Lessons Learned from Accidental Exposure in Radiotherapy. Safety Report No. 17. International Atomic Energy Agency, Vienna.

IAEA, 2001. Investigation of an Accidental Exposure of Radiotherapy Patients in Panamá. International Atomic Energy Agency, Vienna.

IAEA, 2006. Case Studies in the Application of Probabilistic Safety Assessment Techniques to Radiation Sources. TECDOC1494. International Atomic Energy Agency, Vienna.

ICRP, 1997. Protection from potential exposure: application to selected radiation sources. ICRP Publication 76. Ann. ICRP 27 (2).

ICRP, 2000. Prevention of accidental exposure of patients undergoing radiation therapy. ICRP Publication 86. Ann. ICRP 30 (3).

Mijnheer, B.J., Battermann, J.J., Wambersie, A., 1987. What degree of accuracy is required and can be achieved in photon and neutron therapy? Radiother. Oncol. 8, 237–252.

Ortiz López, P., Duménigo, C., Ramírez, M.L., et al., 2008a. Risk analysis methods: their importance for the safety assessment of radiotherapy. Annual Congress of the European Society of Therapeutic Radiology and Oncology (ESTRO 27), 14–17 September 2008, Goteborg. Book of Abstracts.

Ortiz López, P., Duménigo, C, Ramírez, M.L., et al., 2008b. Radiation safety assessment of cobalt 60 external beam radiation therapy using the risk-matrix method. XII Congress of the International Association of Radiation Protection, IRPA 12, 19–24 October 2008, Buenos Aires. Book of Abstracts. Full paper available at: http://www.irpa12.org.ar/fullpaper_list.php.

Rath, F., 2008. Tools for developing a quality management program: proactive tools (process mapping, value stream mapping, fault tree analysis and failure mode and effects analysis). Int. J. Radiat. Oncol. Biol. Phys. 70, S187–S190.

Royal College of Radiologists, Society and College of Radiographers, Institute of Physics and Engineering in Medicine, National Patient Safety Agency, British Institute of Radiology, 2008. Towards Safer Radiotherapy. Royal College of Radiologists, London. Available at: https://www.rcr.ac.uk/docs/oncology/pdf/Towards_saferRT_final.pdf.

Stamatis, D.H., 1995. Failure Modes and Effects Analysis. American Society for Quality Control, Milwaukee, WI.

Thomadsen, B., Lin, S-W., Laemmrich, P., et al., 2003. Analysis of treatment delivery errors in brachytherapy using formal risk analysis techniques. Int. J. Radiat. Oncol. Biol. Phys. 57, 1492–1508.

US Department of Energy, 1996. Hazard and Barrier Analysis Guidance Document. EH-33. Office of Operating Experience Analysis and Feedback, US Department of Energy, Washington, DC.

Vilaragut Llanes, J.J., Ferro Fernández, R., Rodríguez Martí, M., et al., 2008. Probabilistic safety assessment of radiation therapy treatment process with an electron linear accelerator for medical uses. XII Congress of the International Association of Radiation Protection, IRPA 12, 19–24 October 2008, Buenos Aires. Book of Abstracts. Full paper available at: http://www.irpa12.org.ar/fullpaper_list.php.

6. CONCLUSIONS AND RECOMMENDATIONS

6.1. General

(198) This section is a summary of the main safety issues identified retrospectively in Sections 2 and 4 on lessons from accidental exposure and near-misses, as well as an anticipative identification of safety implications of new technologies given in Section 3 and on systematic prospective safety assessments explained in Section 5.

(199) The following conclusion for conventional radiation therapy from ICRP *Publication 86* (ICRP, 2000) is equally applicable, and even more relevant and important, for new technologies: 'purchasing new equipment without a concomitant effort on education and training and on a programme of quality assurance is dangerous'.

(200) Increasingly complex new technologies require a safety strategy that combines:

- Initiatives from manufacturers to incorporate, in their equipment, effective safety interlocks, alerts and warnings, self-test capabilities, and easy-to-understand user interfaces in a language comprehensible by the user. International standards must be adhered to in order to ensure compatibility between equipment from different manufacturers. All these safety measures are applicable to hardware as well as software.
- Revisiting training at three levels: (1) generic training on the in-depth understanding of the science involved in the new technology at both clinical and physical levels, (2) specific training in the equipment and techniques to be used, and (3) 'hands-on' training to obtain the necessary competence before being allowed to use the new techniques in the clinical environment.
- Risk-informed approaches for selecting and developing quality control tests and checks, through the application of prospective methods of risk assessment, to be performed in co-operation with manufacturers.

6.2. Justification of and smooth transition to new technologies

(201) The decision to embark upon a new technology for radiation therapy should be based on a thorough evaluation of expected benefits, rather than being driven by the technology itself. It would be unreasonable to use costly, time-consuming, and labour-intensive techniques for treatments for which the same results could be obtained with conventional, less sophisticated techniques which can be used with confidence and safety.

(202) During technology upgrades, a smooth, step-by-step approach should be followed; for example, moving from conventional to conformal therapy with MLCs through 3D treatment planning to finally arrive at IMRT. Failure to adopt a gradual approach may not only lead to a waste of resources but may also increase the likelihood of accidental exposures.

6.3. Changes in processes and workload

(203) The considerable changes in processes, procedures, tasks, and allocation of staff entailed in the introduction of a new technology need to be planned, commissioned, and quality controlled on a regular basis. The full potential impact of these changes should be assessed.

6.4. Availability and dedication of trained staff

(204) Major safety issues in the introduction of new technologies include the danger of underestimating staff resources, and replacing proper training with a short briefing or demonstration from which important safety implications of new techniques cannot be fully appreciated.

(205) Certain tasks, such as complex treatment planning and pretreatment verification for IMRT, require a substantial increase in resource allocation. The re-assessment of staff requirements, in terms of training and number of professionals, is essential when moving to new technologies.

(206) To prevent shortages of staff with key roles in safety, such as radiation oncologists, medical physicists, and technologists, governments should make provisions for an appropriate system of education and training (in the country or abroad) and have in place a process of certification. In particular, medical physicists, whose activities have a major impact on avoiding catastrophic accidental exposures (e.g. calibration, dosimetry, and physical aspects of quality control), should be integrated as health professionals, and plans should be developed to retain staff who are essential to safety.

(207) Technologists should be involved, together with radiation oncologists and medical physicists, in the decision processes, because technical solutions to monitor patient set-up will become ever more widely available (e.g. image-guided radiation therapy or adaptive radiotherapy).

6.5. Responsibilities of manufacturers and users for safety

(208) Hospital administrators, heads of radiation therapy departments, and staff should remain cognisant of the fact that the primary responsibility for the safe application of new and existing treatment strategies remains with the user. This responsibility includes investigating discrepancies in dose measurements for beam calibration before applying the beam to patient treatments.

(209) Manufacturers should be aware of their responsibility for delivering the correct equipment with the correct calibration files and accompanying documents. They also have a responsibility for supplying correct information and advice, upon request, from the hospital staff. In particular, they should have policies and procedures in place for assisting users to clarify questions on discrepancies in absorbed dose. They should also identify any limitations in performance of their equipment, and pathways which may lead to the misuse of their equipment.

(210) Manufacturers should collect updated information on safety-related operational experience, and disseminate this information rapidly to users (e.g. as safety information bulletins). This dissemination is particularly critical during the introduction of new techniques and technologies, and especially for problems that appear rarely. For example, serious problems may occur when certain conditions happen to coincide; such a coincidence may not be identified during commissioning and subsequent quality control tests.

(211) Programmes for purchasing, acceptance testing, and commissioning should not only address treatment machines but also increasingly complex TPSs, RTISs, imaging equipment used for radiation therapy, software, procedures, and entire clinical processes.

(212) Professional bodies and international organisations should develop codes of practice, and protocols for calibration of specific beam conditions found in new technologies, such as small field size and the absence of charged particle equilibrium.

(213) There is a need to recommission devices and processes after equipment modifications, and software upgrades and updates.

6.6. Dose escalation

(214) Tumour dose escalation requires a reduction of geometrical margins in order to avoid an increase in the probability of complications in normal tissue. Such a reduction is only feasible with an improvement in dose conformality, accompanied by effective immobilisation with accurate and precise patient positioning based on image guidance. Dose escalation also requires a clear understanding of the overall positioning accuracy achievable in clinical practice as a prerequisite to safe margin reduction. Without these features, tumour dose escalation could lead to severe patient complications.

6.7. Radiation doses from increased use of imaging

(215) When making increased use of imaging for simulation, verification, and correction of patient set-up during the course of treatment delivery, an assessment of the additional radiation doses from imaging is necessary for integration of these doses into treatment planning and delivery.

6.8. Omnipresence of computers

(216) Equipment instructions and human–machine communication should be understandable by the users. Procedures should be in place to deal with situations created by computer crashes, which may cause a loss of data integrity. These procedures should include a systematic verification of data integrity after a computer crash during data processing or data transfer.

(217) When introducing an RTIS, it is necessary to develop procedures and to plan commissioning and 'probing' periods to confirm that such a system can be used safely.

6.9. Tests that are no longer effective

(218) When conventional tests and checks are not applicable or not effective for new technologies, the safety philosophy should aim to find measures to maintain the required level of safety. This requirement may lead to the design of new tests or the modification and validation of the old tests. Conscious efforts are required in this regard to avoid compromising safety.

6.10. Consistency in prescription

(219) Protocols for prescription, reporting, and recording, such as those included in ICRU reports, should be kept updated to reflect and accommodate new technologies. Such protocols should be adopted at a national level with the help of professional bodies.

6.11. Co-ordinates, reference marks, and tattoos

(220) Procedures for virtual simulation, and their implications for the whole treatment chain, should be introduced with sufficient training to ensure that the staff are familiar with them and aware of all the critical aspects. A consistent co-ordinate system is required for the whole process from virtual simulation through treatment planning to delivery.

6.12. Handling of images

(221) Written instructions should be visibly posted and followed by the imaging staff who perform the imaging for radiation therapy treatment planning and delivery. These instructions should include procedures for verifying left and right in critical images (e.g. by using fiducial markers), for recording image orientation with respect to the patient, and for ensuring consistency through the whole process from prescription to delivery.

(222) Procedures are also required for selecting the correct images and correct regions of interest, and for deriving electron density from CT, giving specific attention to possible image artifacts and potential geometric distortion.

6.13. Uniformity and clarity in data transfer approaches

(223) When several methods and different protocols for data transfer are used for treating patients in a given department, the patient categories to which the different protocols are applicable should be clearly defined and communicated, including details about which planning system and which data transfer method is applicable.

6.14. Safe interdisciplinary communication

(224) Communication should follow a stated structure regarding content and format, and include formal recording of safety critical issues. Unambiguous communication is essential, especially considering the complexity of radiotherapy and the multidisciplinary nature of the healthcare environment.

6.15. Maintenance, repairs, and notification of the physicist

(225) Procedures to notify a physicist of maintenance or repair activities have been identified as crucial in conventional technology. However, they are even more necessary with new complex technologies, in which modifications, software updates, adjustments, and calibration files can be introduced into the computer dialogue between the various devices, and these might go undetected in the absence of formal notification.

6.16. Prospective safety assessment for selecting quality control checks

(226) The programme of checks should be rationalised and simplified, with the help of manufacturers, by designing proper alerts and warnings, self-test routines especially related to software, easy-to-understand user interfaces, and internal safety interlocks. These measures should be augmented by training in the proper and cautious use of the equipment.

(227) Increased complexity requires a strategy to choose quality control checks based on selective, risk-informed approaches to identify and prioritise tests. In co-operation with manufacturers, mechanisms should be found to perform prospective safety assessments when a new product, technology, or technique is being introduced.

(228) Timely and effective sharing of operational experience is crucial when introducing new techniques and technologies. This could be achieved by organised and structured sharing mechanisms; for example, through the creation of moderated electronic networks and by the early establishment of panels of experts.

6.17. Safety culture

(229) Hospital administrators and heads of radiation therapy departments should provide a work environment that encourages 'working with awareness', facilitates concentration, and avoids distraction. They should monitor compliance with procedures of the quality control programme, not only for the initial treatment plan but also for treatment modifications.

6.18. Reference

ICRP, 2000. Prevention of accidental exposures to patients undergoing radiation therapy. ICRP Publication 86. Ann. ICRP 30 (3).

ANNEX A. SHORT REPORTS ON INCIDENTS WITHOUT SEVERE CONSEQUENCES FROM THE RADIATION ONCOLOGY SAFETY INFORMATION SYSTEM

(A1) This annex is a supplement to Section 4 in the main text. It contains incidents without severe consequences reported to ROSIS, a voluntary web-based safety reporting system for radiotherapy (http://www.rosis.info). The reports provide different examples of errors with new technologies or techniques, and the paragraphs are summarised interpretations of the reports. Although they are not accidental exposures, the reports also contain lessons that can be used to prevent accidental exposures elsewhere.

A.1. Reports relating to R&V systems

A.1.1. ROSIS Incident Report #19: capturing treatment parameters incorrectly on a treatment unit

(A2) Treatment field parameters were transferred from the R&V system to the linear accelerator, including the MUs for dynamically wedged fields. The field size was intentionally modified manually for a treatment field by using the linear accelerator hand control. When the new field size was captured by the R&V system, the previous information on MUs for the dynamic wedge was lost. Two subsequent treatments were given without the dynamic wedge before the error was detected.

A.1.2. ROSIS Incident Report #107: R&V failure to register a given treatment

(A3) A network communication failure occurred between a linear accelerator and an R&V system causing communications to break down. The patient had already been given 30 MU (wedged field) when this failure occurred. To restart communication, the R&V system was rebooted, after which the R&V did not acknowledge that the patient had already received part of the treatment. When treatment resumed, the radiation therapists gave the full treatment, including the previously given 30 MU.

A.1.3. ROSIS Incident Report #116: incorrect MU registration by R&V system

(A4) A patient was treated with a field that was open for one part of the treatment and wedged for another part (using a motorised wedge). When the wedge automatically moved out of the field in order for the open field to be delivered, no information was received by the R&V system, thus continuing to register MUs for the wedged field. A faulty microswitch stopped the correct information from being sent to the R&V system, but the actual treatment was performed correctly.

A.1.4. ROSIS Incident Report #141: error in manual set-up due to R&V system not being used because of 'millennium bug' problems

(A5) An R&V system was taken out of clinical use in a hospital because it was considered that there were millennium bug or year 2000 (Y2K) problems. The system was not replaced on the linear accelerator from which this system had been removed, leading to manual treatment set-up and selection of treatment parameters and accessories. A patient treated with two tangential beams for breast cancer noted that the number of wedges used was different from one day to the next, and asked the radiation therapist if there was a reason for this and if it had any importance. A wedge had been forgotten at the manual set-up of one of the tangential fields of the patient, causing an incorrect absorbed dose to be delivered.

A.1.5. ROSIS Incident Report #690: inadvertent rotational treatment of a patient

(A6) During the first treatment of a patient with an electron field, it was noted that the gantry started to rotate. The prescription was for static treatment, not rotational. An error had been made when preparing the R&V entry of the treatment, where a checkbox had been accidentally checked for rotational treatment. It was also noted in another report to ROSIS (Incident Report #689) that, for this particular type of R&V system, the checkbox for rotational treatment on the screen was placed near the icon for closing the window after finalising the R&V entry, leading to inadvertent activation of rotational treatment.

A.1.6. ROSIS Incident Report #725: problems in selecting the correct field in the R&V system

(A7) A patient was treated with overlapping large and small fields, to be delivered every other day. This was programmed into the treatment schedule of the patient in the R&V system. One day when the patient was treated, it was found that the system allowed the selection and treatment of both series (large and small fields) on the same day, and that after having irradiated the patient with the first field, the R&V system automatically chose the next field with the lowest number, which should have been used for treatment the next day. This was noted after a few MUs had been given. The treatment was interrupted.

A.2. Reports relating to soft wedges on linear accelerators

A.2.1. ROSIS Incident Report #20: treatment with soft wedges in the wrong direction

(A8) During the treatment planning process, the field names of two tangential breast fields (e.g. left medial and left lateral tangential fields) were reversed, thus the treatment parameters were associated with the wrong field names and vice versa.

At treatment set-up, the fields were called up on the R&V system. The technologists set up the correct gantry angle, which did not match the angle recorded in the R&V system. Subsequently, they over-rode the gantry angle on the R&V record, leaving the remaining parameters from the other field. These parameters included dynamic wedge data, thus the wedge direction was from the other field (i.e. the opposite direction to that intended).

A.2.2. ROSIS Incident Report #284: inadvertent loss of wedge code information

(A9) Due to the breakdown of a linear accelerator, a patient was moved to another accelerator for a single fraction. As an inherent part of the design of the R&V system, the wedge information in the R&V system was not transferred automatically to the new treatment unit. The wedge code was manually input properly for the single fraction at the second unit, but when the patient was transferred back to the original unit, the wedge code was not put in again. As a result, the patient received treatment without wedges for three fractions before discovery, causing accidental delivery of the incorrect absorbed dose and dose distribution.

A.2.3. ROSIS Incident Report #310: wrong manual transfer of data on soft wedges into R&V system

(A10) After a patient's treatment had been replanned, the wedge code for the dynamic wedges was not manually entered into the R&V system. This meant that both fields of the treatment set-up were used without the intended wedges for three fractions before discovery.

A.2.4. ROSIS Incident Report #314: incorrect manual entry of soft wedge direction

(A11) The wedge code for a dynamic wedge was entered manually into the R&V system. When performing this entry, the wrong wedge direction was chosen (i.e. 'out' instead of 'in'), leading to incorrect dose distribution in the patient for one fraction before the error was discovered.

A.3. Reports relating to MLCs on linear accelerators

A.3.1. ROSIS Incident Report #132: connectivity problems between R&V system and MLC unit

(A12) For a specific combination of R&V system model and linear accelerator model, there was no verification of the MLC configuration of the treatment fields, i.e. the MLC files containing the information on the MLC settings for each field had to be opened separately on the linear accelerator control software in order to set the MLC configuration, without the possibility of having the correctness of the MLC setting verified. For one field, it was forgotten to open the corresponding

MLC file and to set the MLC configuration, thus starting to irradiate the patient without MLC shielding. At the reported instance, the error was detected after a few MUs and the treatment was stopped, but the incident report indicates that this problem is a recurring issue.

A.3.2. ROSIS Incident Report #707: loss of MLC shape after portal imaging

(A13) The imaging feature in an R&V system was used to make a double-exposed portal image, starting with an open large field in order to make anatomical landmarks more clearly visible in the portal image. The second exposure of the sequence was intended to be of the actual treatment field. However, the MLC settings did not return to the intended treatment settings. This was noticed at the treatment unit and treatment was interrupted.

A.4. Reports relating to computerised TPS tools

A.4.1. ROSIS Incident Report #326: printout of BEV put together incorrectly

(A14) Beam's eye view (BEV) printouts were used to check the shape of the irradiation fields prior to patient treatment by placing the scaled printouts on the treatment couch at a certain distance from the source, and comparing with the light field of the corresponding treatment field. The use of large fields made it necessary to print out the BEV on two sheets of paper and to put them together in order to cover the whole field. When checking a particular field, it was noticed that there was no agreement between the BEV and the light field. When investigating this further, it was found that the two sheets of paper had been put together incorrectly. It was felt that a contributing factor to this mistake was the insufficient identification markings on the BEV printouts by the system.

A.4.2. ROSIS Incident Report #471: transfer of the wrong DRRs of the patient

(A15) At the first treatment of a patient, electronic portal images of the treatment fields were taken. The radiographers on the treatment unit noticed large discrepancies between these images and the DRR images that were used as reference images of the intended field placement. Further investigations revealed that DRR images from a different treatment plan of the same patient had been sent.

A.4.3. ROSIS Incident Report #623: incorrect labelling of simulator film leading to incorrect BEV and incorrect block positioning

(A16) A simulator film showing the intended field shape was labelled on the wrong side. This meant that when digitised into the TPS, the BEV was mirrored in relation to the intended field shape, leading to the incorrect positioning of the lead shielding in the treatment field. Since the BEV was used to verify the correctness of the block positioning before treatment, the patient was treated with the block in the incorrect position before the error was discovered through portal imaging.

A.5. Reports relating to imaging for treatment planning

A.5.1. ROSIS Incident Report #454: CT images associated with the wrong patient when entered into the TPS

(A17) When performing an electronic transfer of CT images, it was necessary to manually associate the data with a specific patient since the patient identity information in the CT data was not recognised electronically by the TPS. When the CT images of one patient were transferred into the TPS, the CT data were introduced into the records of another patient. The error was detected at a later point in the treatment planning process.

A.6. Reports relating to virtual simulation

A.6.1. ROSIS Incident Report #161: problems due to inadvertent energy selection originating in the virtual simulation process

(A18) When performing virtual simulation with a particular system, a field had to be entered electronically into the patient CT data in order to set an isocentre in the simulation process. Thus, a specific photon energy had to be selected for the field despite the fact that the staff did not know which energy was the most appropriate at that time in the process. As a rule, 6 MV was always chosen in the clinic when creating the field for setting the isocentre. When creating the treatment plan for a pelvic treatment, the planner should change this initial energy to a higher energy, but this was not done for a particular patient. The treatment was a three-field technique with two wedged lateral fields. Prior to starting treatment delivery, it was noted that the energy was too low for a pelvic treatment, and the planner made a new plan with a higher photon energy. When the energy of the field was changed, the information on the wedges of the lateral fields was lost. This was not noticed by the treatment planner. Furthermore, the MUs were already calculated and checked before the change of energy and were not rechecked after the change. The mistake was discovered after the first field was given.

A.6.2. ROSIS Incident Report #573: different length units in virtual simulation and linear accelerator

(A19) A very small field (6 mm) was virtually simulated. This treatment was not calculated by the computerised TPS but was calculated manually, and therefore it was not electronically transferred. When the radiographers on the treatment unit were recording the treatment parameters into the R&V system, they interpreted the field size incorrectly. While the length unit used in the virtual simulation system was millimetres, the different length units on the linear accelerator led to the field size being interpreted as 0.6 mm. The mistake was discovered before the start of treatment.

Annals of the ICRP

Published on behalf of the International Commission on Radiological Protection

Aims and Scope

The International Commission on Radiological Protection (ICRP) is the primary body in protection against ionising radiation. ICRP is a registered charity and is thus an independent non-governmental organisation created by the 1928 International Congress of Radiology to advance for the public benefit the science of radiological protection. The ICRP provides recommendations and guidance on protection against the risks associated with ionising radiation, from artificial sources widely used in medicine, general industry and nuclear enterprises, and from naturally occurring sources. These reports and recommendations are published approximately four times each year on behalf of the ICRP as the journal *Annals of the ICRP*. Each issue provides in-depth coverage of a specific subject area.

Subscribers to the journal receive each new report as soon as it appears so that they are kept up to date on the latest developments in this important field. While many subscribers prefer to acquire a complete set of ICRP reports and recommendations, single issues of the journal are also available separately for those individuals and organizations needing a single report covering their own field of interest. Please order through your bookseller, subscription agent, or direct from the publisher.

ICRP is composed of a Main Commission, a Scientific Secretariat, and five standing Committees on: radiation effects, doses from radiation exposure, protection in medicine, the application of ICRP recommendations, and protection of the environment. The Main Commission consists of a Chair and twelve other members. Committees typically comprise 10–15 members. Biologists and medical doctors dominate the current membership; physicists are also well represented.

ICRP uses Working Parties to develop ideas and Task Groups to prepare its reports. A Task Group is usually chaired by an ICRP Committee member and usually contains a number of specialists from outside ICRP. Thus, ICRP is an independent international network of specialists in various fields of radiological protection. At any one time, about one hundred eminent scientists and policy makers are actively involved in the work of ICRP. The Task Groups are assigned the responsibility for drafting documents on various subjects, which are reviewed and finally approved by the Main Commission. These documents are then published as the *Annals of the ICRP*.

International Commission on Radiological Protection

Scientific Secretary: **C.H. Clement**, *ICRP, Ottawa, Ontario, Canada; sci.sec@icrp.org*

Chair: **Dr. C. Cousins**, *Department of Radiology, Addenbrooke's Hospital, Cambridge, UK*

Vice-Chair: **Dr. A.J. González**, *Argentina Nuclear Regulatory Authority, Buenos Aires, Argentina*

Members of the 2009–2013 Main Commission of the ICRP

J.D. Boice Jr, *Rockville, MD, USA*
J.R. Cooper, *Didcot, UK*
J. Lee, *Seoul, Korea*
J. Lochard, *Fontenay-Aux-Roses, France*
H.-G. Menzel, *Genève, Switzerland*
O. Niwa, *Chiba, Japan*
Z. Pan, *Beijing, China*

R.J. Pentreath, *Cornwall, UK*
R.J. Preston, *Research Triangle Park, NC, USA*
N. Shandala, *Moscow, Russia*
E. Vañó, *Madrid, Spain*

Emeritus Members
R.H. Clarke, *Hampshire, UK*
B. Lindell, *Stockholm, Sweden*
C.D. Meinhold, *Brookhaven, NY, USA*
F.A. Mettler Jr., *Albuqverqve, NM, USA*
W.K. Sinclair, *Escondido, CA, USA*
C. Streffer, *Essen, Germany*

The membership of the Task Group that prepared this report was:

P. Ortiz López
J.-M. Cosset
P. Dunscombe

O. Holmberg
J.-C. Rosenwald
L. Pinillos Ashton

J.J. Vilaragut Llanes
S. Vatnitsky